RAT CATCHING

FOR THE USE OF SCHOOLS

RAT CATCHING
FOR THE USE OF SCHOOLS

H. C. BARKLEY

A new, annotated edition of a book first published in 1896
under the title *Studies in the Art of Rat Catching*, and with a
new Introduction and annotations, by B.J. Coman, author of
Tooth and Nail: *The Story of the Rabbit in Australia*.

Connor Court Publishing

Text by H.C. Barkley, first published in 1896.

Connor Court Publishing Pty Ltd
PO BOX 7257
Redland Bay QLD 4165

sales@connorcourt.com
www.connorcourt.com

ISBN: 978-1-925501-00-1

Front Cover: Istockphoto

Printed in Australia

PREFACE TO THE ORIGINAL EDITION

My publisher writes to say that he, and he thinks others too, would like to know how I ever came to write such a book as this! It came about in this way. Some two years ago, I was about to leave England for a considerable time, and a few days before starting, I went to stay in a country house, full of lads and lassies, to say good-bye. One evening, while sitting over the study fire, the subject of rat-catching came up, and, as the aged are somewhat wont to do, I babbled on about past days and various rat-catching experiences, till one of the boys exclaimed, "I say, what sport it would be if they would only teach rat-catching at school! Wouldn't I just work hard then, that's all!"

The stories came to an end at bed-time, and I was then pressed by my hearers to write from foreign lands some more of my old reminiscences, and I readily gave a promise to do so. In this way most of the following stories were written; and in writing them, I endeavoured to carry out the idea that they were exercises to be used in schools.

I don't anticipate that head-masters will very generally adopt the book in their schools; but I hope it may, in some few instances, give boys a taste for a wholesome country pastime.

The characters and incidents are rough, very rough, pen and ink sketches of real people and scenes, and the dogs are all dear friends of past days.

CONTENTS

INTRODUCTION TO THE NEW EDITION

This present book is one of an intended series dealing with topics which, today, would almost certainly meet with a great deal of tongue clucking by the chattering classes. Not least amongst the 'taboo' subjects these days is the matter of hunting. That a book should be published (or republished) in these enlightened times of ours which actually recommends the hunting and destruction of a wild animal as a useful trade to be taught to young children, is quite beyond the pale. Now, of course, if those young children happen to be the offspring of indigenous hunter-gatherers anywhere in the world, then the situation is quite different, and they will be encouraged to take up their 'traditional ways'. But for children of European descent, the very thought of using one partly domesticated animal to hunt another for either food, monetary gain, or just simple enjoyment, is regarded as a regression to barbarism. Better by far that they should do their hunting in those super-realistic computer games, where vanquished foes—the bloodied corpses of fellow human beings—can be simply erased with a single click of the mouse. In this way, does our Civilisation advance! Later in their careers they may very well learn other skills of hunting and killing as, for instance, in the stock exchange or other commercial settings. And who is to say which is more barbaric and destructive of human character, the old form of hunting or the new?

H.C. Barkley's little manual on rat catching (and rabbit catching) is from a past era when life was much simpler and our propensity to call a pest a pest was not diminished by the selective fads and niceties of modernity. Most important of all, it was an era when the Aristotelian idea of ends—the proper conduct of a life towards a desired conclusion—was of importance not just for humans, but for all living things. It was proper and ordained for rats to live rat-centred lives, doing the things rats do, just as it was proper for ferrets to do the things that come natural to them. And, as Barkley himself indicates in his Introduction, it was ordained in the great scheme of things that dogs should come to their human masters and not the other way round, for this, too, represents a natural order in the hierarchy of being. It was proper also, that each human occupation should be regarded as having an intrinsic worth and the value of the trade of rat-catching should be seen in no less a light than that of any other profession. Indeed, as Barkley hints in the book, the value of a rat-catcher might well be regarded in a higher light than that of your average politician who, as Barkley reminds us, is prone to giving speeches when Parliament is not sitting or, indeed, at every possible opportunity! And this, as Barkley tells us, is why he decided on a school text book for prospective rat catchers.

But what of Henry Barkley himself? In fact, we know little about him. He was born about 1825 and he died some 75 years later. By trade, he was a civil engineer and we know that, together with his elder brother Trevor, he planned and built a railway linking the Danube to the Black Sea port of Constanta. We know, too, that he was something of a traveller, publishing an account of a journey through Asia Minor and Armenia (*A Ride Through Asia Minor and Armenia*, published 1891), plus two other books dealing with his experiences in Bulgaria. His only other publication known to this writer was *My Boyhood,* published in 1877.

This, after all, was the age of the intrepid English explorer and traveller, risking all for 'progress', personal curiosity, and the glory of the Empire. Barkley was very probably moved to give an account of his travels by the enthusiastic response given to earlier books in this genre—for example, A.W. Kinglake's *Eothen*; or *Traces of Travel Brought Home from the East* (1844), and Fred Burnaby's *A Ride to Khiva* (1876), and *On Horseback through Asia Minor* (1877).

Whatever the literary merits of Barkley's travel books, there is no question that his account of his childhood, *My Boyhood*, and this present volume on rat catching establish him as a first rate author for children. In the Preface to the first edition of his rat-catching book (1896), he recounts how young people who had listened to some of his accounts of his childhood pressed him to enlarge on the subject. This he did, apparently while he was overseas, and *Rat Catching* was the happy result. For the interested reader who may wish to read *My Boyhood*, I should say that a portion of the material in the rat-catching book also appears in *My Boyhood*, especially in relation to dogs. Barkley had a great fondness for dogs, and I have every reason to believe that the dogs in his books were real animals owned by him as a boy. Indeed, his two children's books appear to be a mixture of both fact and fiction or, perhaps more accurately, fact and embellished fact.

Now, I have indicated above that *Rat Catching* is a children's book. This is not quite true. It is, rather a book for young boys and old men. I make no apologies here for sexist tendencies, for the truth of the matter is that young girls are almost always aversive to both rats and ferrets. When you read Barkley's book, you will see that he agrees with me. Young girls far prefer ponies and, maybe, pet rabbits. Can you imagine a roomful of pink plastic toys all marked My Little Ferret? The idea is ludicrous. Now I include old men in my proposed readership because, if they are like me, this book will rekindle a thousand fond memories of childhood. I too had ferrets, as did nearly all the young boys in our farming district. If (like me again) you are in the 'senior citizen' category (a weasel term for 'old codger') and were brought up in rabbit country anywhere in southern Australia, you will be astounded by this book. You will be astounded that someone else, living in a faraway country, should have had the very same experiences with dogs and ferrets as you did, and that he should offer the very same advice that you yourself would give to your grandson. It leaves you with a lump in your throat, and you find yourself fighting back the tears as fond memories of champion rabbit dogs or of particularly successful hauls of rabbits come flooding back.

But much more is going on in this book. From the very title, you immediately realise that Barkley is poking a bit of fun at the educational arrangements of his day. Indeed, he would invert the prevailing school set-up of his day by putting practical, agricultural pursuits well ahead of 'academic' subjects like Latin and Greek. Here of course, his young readers will be right on side! What young boy would not prefer a day out in the fields with dogs and ferrets to the boring classroom and lessons on Cicero? Immediately then, he establishes a rapport with the young reader.

Sadly, all that Barkley describes belongs to a bygone age. The era of ferreting rabbits in Australia has now all but passed. A few young boys still keep ferrets for rabbiting, but the numbers are dwindling. Indeed, in many southern areas of the Continent, rabbit numbers have declined markedly due to the success of biological control agents and to profound changes in land management practices. Country boys who might once have kept ferrets probably ride dirt bikes these days. Indeed, the keeping of ferrets has probably now shifted to the cities where, today, increasing numbers of people, and not all of them young people, keep ferrets as pets. But these are weak, emasculate creatures (the ferrets I mean) often mutilated to remove their anal scent glands. They are destined to live sad lives inside houses, to be walked on a lead, to smell of sweet and sickly unguents, to suffer the indignity of being bathed regularly, and to experience totally unfulfilled lives as exhibition pieces. Aristotelian teleology has been savagely cast aside in the interests of human fashion.

To return to the book before us: even if you, as an older reader, have no background in rabbiting, the book will still enthral you because, behind the matter-of-fact account of keeping ferrets and dogs, etc., there lurks a very sharp wit and a first rate storyteller. The man who wrote this book not only has an intimate and first-hand knowledge of his subject, he also has that rare ability to return at will to his own childhood and to re-create the past as a living thing, "The Child is father of the Man", as Wordsworth says. This is the sort of book a father should read to his son at bedtime, to the mutual advantage of both reader and listener. All good children's books are of this nature.

It is also very evident to the reader that Barkley has a deep love of the natural world around him and an ability to vividly portray the sights, sounds and smells of the agricultural world around him. He is a keen observer of fine details. There are passages in the book which will remind one of a Gilbert White or a Richard Jeffries. We get from Barkley's description a very good idea of the physical and the social circumstances of village life in late 19C England.

Now it so happens that the publisher of this new edition of Barkley's book, Connor Court Publishing, had its headquarters at Ballarat in Victoria, not all that far from where wild rabbits were first released in Australia some 160 years ago. It seems fitting then, that I should attempt to place Barkley's account in an Australian setting, explaining just how a book, published in England long before the Australian colonies had any experience of European rabbits (but they did have rats!), has a huge relevance to our own history.

Recall firstly that, whilst Barkley spends a good deal of time describing ratting, he has two chapters on rabbit catching. A good deal of the content here is directly relevant to the use of ferrets in Australia. Moreover, the general advice on ratting dogs applies also to rabbiting dogs, especially those used in ferreting. For instance, that marvellous piece of advice concerning choice of dog ("the shorter the pedigree, the better the dog") is absolutely true of rabbiting dogs in Australia as well as ratting dogs in England. It goes without saying that all of the advice on keeping ferrets is pertinent to the Australian scene.

Here in Australia, we rarely use ferrets and dogs to hunt rats. Sometimes, of course, rats may occupy parts of a rabbit burrow, and ferrets will kill them or drive them out. But this is an unintended consequence, something that military people call collateral damage. The nature of our farming, of our soils and of our inland climate means that rats are not of the same agricultural importance or general nuisance value as they are in England and continental Europe. We are much more concerned with plague mice and, especially, with the control of rabbits. There are a few other small details in Barkley's book which have limited relevance in Australia. Historically, we have seldom used long-netting as a method of rabbit capture. I will let Barkley explain the technique for you (Chapter 8). Suffice to say that, for most of our rabbit country, the technique would have no application. However, I must add that the technique is still occasionally used in urban situations where rabbit control by conventional means (poisoning, shooting, fumigating, etc.) is not possible. Also, it was once widely used for the live capture of hares in the bad old days when they were used for greyhound racing—a sport far removed from hunting. Here and there in Australia you may still come across a country lane called 'Plumpton Road'—this indicates that a greyhound racing course was once in the vicinity. Hares, by the way, cannot be captured by ferreting as they do not inhabit burrows.

In Australia, we never use lines on ferrets either to trace their underground movements in burrows, or to draw them out after an unintended "kill". Our warrens (for that is the name we give to rabbit burrows out here, while in England the word is used to mean simply an enclosure for rabbits) are far too large on average, and the tunnels tortuous. In sandy country especially, the tunnels of warrens may have many sharp turns and intersections and may even be storied—that is to say, criss-crossing each other at different levels. A ferret with a cord attached might easily become snagged and, in any case, dragging a huge length of line through a long burrow would be like running a marathon with a bag of spuds on your shoulder. The heavier soils of England are a different matter. The burrows will be shorter and shallower.

Like an English rat-catcher, an Australian rabbiter could once eke out a living by plying his trade, but in very different circumstances. Obviously, our English counterpart could not sell his quarry for human food but, in Australia, ferreters could earn quite good money by selling rabbit carcases and rabbit skins. As an example, my older brother had by the time of his marriage (in his mid-twenties) saved up sufficient funds from weekend ferreting and the subsequent sale of rabbits and rabbit skins (used to make felt hats), to supply a good part of the finance needed for a new house. According to Barkley, an English ratter of his day (circa 1850, say) earned twopence per carcase. I suspect that an Australian boy, circa 1950, would have received the equivalent of at least 50 cents (modern value) per rabbit carcase. A good day's ferreting might yield 40 rabbits or more, so this was very good pocket money indeed. In my boyhood, the few professional rabbiters I knew usually combined ferreting and trapping to supply themselves with a living. Later, professional rabbiters turned to night shooting using spotlights and much larger hauls were obtained. Even so, I know of no rich rabbiters!

I should end this short introduction by returning to the matter of animal ethics and the modern aversion to killing animals (except cockroaches, flies, mice, non-native rats and European carp!), and to the whole idea of introducing young boys to a 'blood sport'. To be fair, a certain aversion to the hunting and slaughter of animals for food is as old as human history itself. In those grisly scenes enacted in the Cyclops cave, Homer reminds us of just what is involved in eating the flesh of another living animal. Hesiod, too, felt that it was improper for humans to devour other animals. In his poem Purifications, Empedocles (5th C. BC) denounces the eating of living creatures as the greatest pollution of mankind. Many centuries later, the Neoplatonist, Porphyry, wrote a whole treatise entitled *On Abstinence from Animal Food*. Coming closer to our own time and to the subject matter of this book—the hunting of rats and rabbits—the reader may be surprised to learn that one of the early official concerns regarding the rabbit pest in the Australian Colonies related not to any physical damage caused by the new pest but, rather, to the brutalisation of young boys who might become accustomed to clubbing the animals to death in 'rabbit drives'.

In spite of such ancient and persistent reservations about the hunting and killing of animals, it remains the case that humans have always hunted and killed animals, mostly for food, but often purely for sport. Homer may have given us an image of the Cyclops devouring Odysseus's men, but his great hero himself had a hunting dog (Argus), as did his son, Telemachus. Likewise, with the advent of Christianity, many of the early Fathers were not in favour of hunting, but this did not prevent Chaucer's monk in the *Canterbury Tales* from taking up the sport:

> He yaf nat of that text a pulled hen,
> That seith that hunters beth nat hooly men.

If I may return once more to my Aristotelian theme of *telos*, it is in the very nature of predators to hunt and of a certain proportion of a prey species to be killed by them. There is also a powerful argument to the effect that humans can never wholly shed their predatory background. Although I have a very poor opinion of evolutionary psychology as a science, I am prepared to believe that the hunting urge remains in us, however submerged by the exigencies of modernity. Those who would criticise young boys for stretching a rabbit's neck, after it has been extricated from a purse net, need to weigh up a number of countervailing arguments.

We accept the rather shaky logic that factory farming is OK

because no animal suffers physical pain. We do so because we want cheap and easily procured food. Aristotle would argue (I feel sure) that the issue lies elsewhere, specifically in that notion of the *telos* or 'end' towards which each living thing aspires. The hunted animal may have its life drawn short, but it is a life at least. It is not just a meat-producing unit. It is not my purpose here to end with a long philosophical discussion of the topic, but I refer the interested reader to the famous work by the philosopher Ortega y Gasset, *Meditations on Hunting* (1972). I might also put to the interested reader the following choice: in the matter of bringing up your son/s, which do you feel will provide the better outcome—a day in the Australian bush interacting with the realities of human existence, or a day in a games arcade learning how to despatch the maximum number of your digitised opponents with the minimum of effort and the maximum of brutality? I had the great good fortune to be given the former option and it provided me a childhood of enormous happiness.

So, I suspect, did it provide similar joy to H.C. Barkley. The proof is in the pages you are about to read. Enjoy them after you have tucked in to your RSPCA-approved chicken dinner from Coles or Woolies!

Brian Coman is author of *Tooth and Nail, the Story of the Rabbit in Australia* [Text Publishing, 1999], and a former research biologist working with pest animals in Australia. He grew up on a farm in rabbit country in central Victoria and shared many of the experiences recounted by H.C. Barkley in this book. He has also published a book of essays, *A Loose Canon* [Connor Court, 2007], and a critique of modernity, *Against the Spirit of the Age* [Connor Court, 2015].

INTRODUCTION TO THE 1896 EDITION

Ever since I was a boy, and ah! long, long before that, I fancy, the one great anxiety of parents of the upper and middle classes blessed with large families has been, "What are we to do with our boys?" and the cry goes on increasing, being intensified by the depreciation in the value of land, and by our distant colonies getting a little overstocked with young gentlemen, who have been banished to them by thousands, to struggle and strive, sink or swim, as fate wills it. At home, all professions are full and everything has been tried; and, go where you will, even the children of the nobles may be found wrestling with those of the middle and working classes for every piece of bread that falls in the gutter. Nothing is *infra dig* that brings in a shilling, and all has been and is being tried. The sons of the great are to be found shoulder to shoulder with "Tommy Atkins," up behind a hansom cab, keeping shops, selling wines, horses, cigars, coal, and generally endeavouring feebly to shoulder the son of the working man out of the race over the ropes. Fortunately, Heaven tempers the wind to the shorn lamb, and I believe it has done so now. I believe kind Dame Nature during the last summer stepped in and opened out an honourable path for many gentlemen's sons, that I think will be their salvation, and at all events, if it does not make them all rich, will, if they only follow it, make them most useful members of society and keep them out of mischief and out of their mammas' snug drawing-rooms. I have followed the path myself, and, after fifty years' tramp down it, have been forced to abandon it owing to gout and rheumatism. I have not picked up a big fortune at it, or become celebrated, except quite locally; but I have had good time and helped the world in general, and am content with my past life.

I was the son of a worthy country parson, who in my youth proposed to me in turn to become a judge, a bishop, a general, a Gladstone, a Nelson, a Sir James Paget, and a ritualistic curate; but when talking to me on the subject the good old man always said, "Mind, my boy, though I propose these various positions for you, yet, if you have any decided preference yourself, I will not thwart you, I will not fly in the face of nature."

For some time I thought I should rather like to be a bishop, and to this day I think I should have made a good one; but the voice spoke at last, and my destiny was settled.

With the modest capital of five shillings given me by my father, and a mongrel terrier, given me by a poacher who had to go into retirement for killing a pheasant and half killing a keeper, I began my career as a—but I had better give you one of my professional cards. Here it is:

BOB JOY,

RAT-CATCHER
To H.R.H. The Prince of Wales,
The Nobility and Gentry

I had a struggle at first. Rats, full-grown ones, only fetched twopence each, and the system adopted by farmers of letting their rat-killing, for, say, three, pounds a year for a farm of 400 acres, almost broke me; but I stuck to my profession, and do not regret having done so.

In those days, and during all my active life, I have had to work to live, owing to the constant scarcity of rats; but if I managed to make a living then, what might not be done now, when Nature has sent the rat to our homesteads by thousands, and farmers and others are being eaten off the face of the earth by them?

Why, my dear friends, your fortune stares you in the face, and you have only to stretch out your hand and grasp it—no! I have made a mistake; you have a little more to do—you have, first, to learn your profession, which is no easy matter; and to enable you to do this, I intend writing the following book for the use of schools (which I herewith dedicate to the Head Masters of Eton, Harrow, Westminster, Rugby, and all other schools); but in placing this book on your school-desk, allow me to say that it is no good having it there through the long school hours unless you open it, read it, and deeply ponder over it; and more, my dear boys, let me pray that you will take it home with you, and, casting aside your usual holiday task, study it well, and, as far as possible, actively put in practice what I am going to try and teach you. Some fathers may wish their sons to enter on a more humble course of life, but this I rather doubt. However, should they do so, it will be only so much the better for those who take it up; there will be more room for them. Most mothers, I fear, will object to it on the ground that rats and ferrets don't smell nice; but this objection is not reasonable. They might as well say that the whiff of a fox on a soft December morning as you ride to covert is not delicious!

Respect your parents, respect even their prejudices; gently point out to your father that you are ambitious and wish for a career in which you can distinguish yourself. Above all, respect your mother, and show your respect by not taking ferrets or dead rats in your pockets into her drawing-room, and by washing your hands a little between fondling them and cuddling her. But to finish this sermon, let me point out that though in this great profession you will be everlastingly mixed up with dogs of all sorts, always make *them* come to *you,* and *never go to them.*

One last word. If in the following pages you come across a bit of grammar or spelling calculated to make a Head Master sit up, excuse it, and remember that I have been a rat-catcher all my life, and as a class we are not quite A1 at book learning.

1

In the following elementary treatise for the use of public schools, I propose following exactly the same plan as my parson (a good fellow not afraid of a ferret or a rat) does with his sermons—that is, divide it into different heads, and then jumble up all the heads with the body, till it becomes as difficult to follow as a rat's hole in a soft bank; and, to begin with, I am going to talk about ferrets, for without them rat-catching won't pay.

Where ferrets first came from I am not sure, but somewhere I have read that they were imported from Morocco, and that they are not natives of Great Britain any more than the ordinary rat is.[1] If they were imported, then that importer ranks in my mind with, but before, Christopher Columbus and all such travellers. Anyhow it is quite clear that nowhere in Great Britain are there wild ferrets, for they are as distinct from the stoat, the mouse-hunter, the pole-cat, etc., as I am from a Red Indian; and yet all belong to the same family, so much so that I have known of a marriage taking place between the ferret and pole-cat, the offspring of which have again married ferrets and in their turn have multiplied and increased, which is a proof that they are not mules, for the children of mules, either in birds or beasts, do not have young ones.

There are two distinct colours in ferrets—one is a rich dark

---

[1] The ferret is actually a domesticated form of the polecat, an animal native to much of Europe. However, Barkley is probably correct in supposing that the domesticated form came from places other than the UK. It has been suggested that domestication of the species first occurred in North Africa. There is good evidence for the use of ferrets by the Romans and specific mention of them is made by Pliny and Strabo. Perhaps, then, it was the Romans who domesticated them. Since ferrets can successfully breed with polecats, there is every reason to suppose that wild hybrids may exist in the UK.

brown and tan, and the other white with pink eyes; and in my opinion one is just as good as the other for work, though by preference I always keep the white ferret, as it is sooner seen if it comes out of a hole and works away down a fence or ditch bottom.[2] I have never known a dark-coloured ferret coming among a litter of white ones or a white among the dark; but there is a cross between the two which produces a grizzly beast, generally bigger than its mother, which I have for many years avoided, though it is much thought of in some parts of the Midlands. I fancy (though I may be wrong) that the cross is a dull slow ferret, wanting in dash and courage, and not so friendly and affectionate as the others, and therefore apt to stick with just its nose out of a hole so that you can't pick it up, or else it will "lay up" and give a lot of trouble digging it out.[3]

For rat-catching the female ferret should always be used, as it is not half the size of the male, and can therefore follow a rat faster and better in narrow holes; in fact, an ordinary female ferret should be able to follow a full-grown rat anywhere. The male ferret should be kept entirely for rabbiting, as he has not to follow down small holes, and being stronger than the female can stand the rough knocking about he often gets from a rabbit better than his wife can. In buying a ferret for work, get one from nine to fifteen months old, as young ferrets I find usually have more courage and dash than an old one. They have not been so often punished and therefore do not think discretion the better part of valour. However this will not be found to be an invariable rule. I have known old ferrets that would have faced a lion and seemed to care nothing about being badly bitten; whereas I have known a young ferret turn out good-for-nothing from having one sharp nip from a rat. Such beasts had better be parted with, for a bad,

---

[2] According to modern literature on the subject (of which there is a huge amount) some white ferrets can have black eyes. The pink-eyed white is often called the 'albino ferret'. Barkley is otherwise correct – they differ is colour only.

[3] See footnote 14 on page 33

slow, or cowardly ferret is vexation of spirit and not profitable.

If I am buying brown ferrets I always pick the darkest, as I fancy they have most dash. This may be only fancy, or it may be the original ferret was white and that the brown is the cross between it and the polecat, and that therefore the darker the ferret, the more like it is in temper as well as colour to its big, strong, wild ancestor. Anyhow I buy the dark ones.

If I am buying female ferrets, I like big long ones, as a small ferret has not weight enough to tackle a big rat, and therefore often gets desperately punished. I like to see the ferrets in a tub, end up, looking well-nourished and strong; and directly I touch the tub I like to see them dash out of their hidden beds in the straw and rush to spring up the sides like a lot of furies. When I put my hand in to take one, I prefer not to be bitten; but yet I have often known a ferret turn out very well that has begun by making its teeth meet through my finger. When I have the ferret in hand, I first look at its tail and then at its feet, and if these are clean it will do. If, on the other hand, I find a thin appearance about the hairs of its tail and a black-looking dust at the roots, the ferret goes back into the tub; or if the underside of the feet are black and the claws encrusted with dirt, I will have nothing to say to it, as it has the mange and will be troublesome to cure.[4] All this done, I put the ferret on the ground and keep picking it up and letting it go; if when I do this its sets up the hairs of its tail, arches its back and hisses at me, I may buy it; but I know, If I do, I shall have to handle it much to get it tame. If, on the other hand, when I play with it the ferret begins to dance sideways and play, I pay down my money and take it at once, for I have never

---

[4] This is the common mange mite, *Sarcoptes scabei*, which can cause problems in many animal species. As well as infecting the ferret's feet (sometimes called 'footrot'), it can also cause skin problems elsewhere on the body. Effective treatments are now available.

known a playful ferret to prove a bad one.[5]

If when you get the ferret it is wild and savage, it should be constantly handled till it is quite tamed before it is used. Little brothers and sisters will be found useful at this. Give them the ferret to play with in an empty or nearly empty barn or shed where it cannot escape. Put into the shed with them some long drain pipes, and tell them to ferret rats out of them. The chances are they will put the ferret through them and pick it up so often, that it will learn there is nothing to fear when it comes out of a real rat's hole, and will ever after "come to hand" readily. You had better not be in the way when the children return to their mother or nurse. I have had disagreeable moments on such occasions.

Having got all your ferrets, the next question is how to keep them. I have tried scores of different houses for them. I have kept them in a big roomy shed in tubs, in boxes, and in pits in the ground; but now I always use a box with three compartments. The left-hand compartment should be the smallest and filled with wheat-straw well packed in, with a small round hole a little way up the division, for the ferrets to use as a door. The middle compartment should be empty and have the floor and front made of wire netting, to allow light, ventilation and drainage. The third compartment should be entered from the middle one by a hole in the division, but should have a strong tin tray fitting over the floor of it covered with sand, which can be drawn out and cleaned; the front of this compartment, too, should be wire netting. The sand tray should be removed and cleaned every day, even Sundays. The house should stand on legs about a foot high. Each compartment should have a separate lid, and the little entrance holes through the division should have a slide to shut them, so that any one division can be opened without all the

---

[5] This behaviour is sometimes called the 'weasel war dance' and is common to some other members of the Mustelidae. Barkley is probably correct in his interpretation of the behaviour as a form of excited play

ferrets rushing out. The bed should be changed once a week. Such a box as I have shown is large enough for ten ferrets.[6] For a mother with a family a much smaller box will suffice, but it should be made on the same plan. For bedding use only wheat-straw. Either barley-straw or hay will give ferrets mange in a few days.[7]

After housing the ferrets, they will require feeding I have always given my ferrets bread and milk once or twice a week, which was placed in flat tins in the middle compartment; but care should be taken to clean out the tins each time, as any old sour milk in them will turn the fresh milk and make the ferrets ill. The natural food of ferrets is flesh—the flesh of small animals—and therefore it should be the chief food given. Small birds, rats and mice are to them dainty morsels, but the ferrets will be sure to drag these in their beds to eat and will leave the skins untouched; these should be removed each day. When my ferrets are not in regular work they are fed just before sunset; if they are fed in the morning they are no good for work all day, and one can never tell (excepts on Sundays) that one of the dogs may not find a rat that wants killing. The day before real work, I give the ferrets bread and milk in the morning, and nothing on the day they go out until their work is over. This makes them keen. Remember ferrets work hard in a big day's ratting, and therefore should be well nourished and strong; a ferret that is not will not have the courage to face a rat.

I have listened to all sorts of theories from old hands about feeding ferrets, but have followed the advice of few. For instance, I have been told that if you give flesh, such as rats and birds, to

---

[6] The dimensions are not given by Barkley, but a structure to house ten ferrets would need to be quite large. Interested readers might consult the internet, where there is a huge volume of literature on the subject due largely to increased interest in ferrets as urban pets.

[7] This seems doubtful as the mange mite has no clear connection with different types of straw.

ferret that has young ones, it will drag it into the straw among the little ones, who will get the blood on them, and then the mother will eat them by mistake. All I can say is, I have reared hundreds of young ferrets and have always given the mothers flesh. It is true that ferrets will eat their young, and the way to bring this about is to disturb the babies in the nest. If you leave them quite alone till they begin to creep about I believe there is no danger.

Then, many old rat-catchers never give a ferret a rat with its tail on, as they believe there is poison in it. I remember one old fellow saying to me as he cut off the tail before putting the rat into the ferrets' box "Bar the tail—I allus bars the tail—there's wenom in the tail." There may be "wenom" in it; but, if there is, it won't hurt the ferrets, for they never eat it or the skin.

If ferrets are properly cared for they are rarely ill, and the only trouble I have ever had is with mange, which, as I have said before, attacks the tail and feet. Most rat-catchers keep a bottle of spirits of tar,[8] with which they dress the affected parts. It cures the mange, but, by the way the poor little beasts hop about after being dressed, I fear it stings dreadfully. I have always used sulphur and lard, and after rubbing it well in a few times I have always found it worked a cure. The objection to sulphur and lard is that it does not hurt, for I have noticed that sort of man generally prefers using a remedy that hurts a lot—that is, where the patient is not himself, but an animal.

No big day's ratting ever takes place without a ferret getting badly bitten. When this is so, the ferret should never be used again until it is quite well. It should be sent home and put in a quiet box, apart from the others, and the bites gently touched

---

[8] I take this to refer to a product known in Australia as Stockholm Tar – a type of pine-tar formerly used to treat shearing wounds in sheep. In the famous Australian Song *Click Go the Shears*, there is reference to a 'tar pot' – something to be found in every Australian shearing shed in former times. Today, there are much more effective treatments.

with a little sweet oil[9] from time to time; or, if it festers much, it should be sponged with warm water.

I have often had ferrets die of their wounds, and these have usually been the best I had. Again, with wounds the old rat-catcher uses the tar-bottle, chiefly, I think, because it hurts the ferret, and therefore must have "a power of virtue."

Before going further I should point out to all students of this ennobling profession that the very first thing they have to learn is to pick up a ferret. Don't grab it by its tail, or hold it by its head as you would a mad bulldog; but take hold of it lightly round the shoulders, with its front legs falling gracefully out below from between your fingers. Then when you go to the box for your ferrets, and they come clambering up the side like a pack of hungry wolves, put your hand straight in among them without a glove, and pick up which one you require. Don't hesitate a moment. Don't dangle your hand over their heads till you can make a dash and catch one. The ferrets will only think your hand is their supper coming and will grab it, with no ill intent; but if you put it down steadily and slowly, they will soon learn you only do so to take them out, and your hand will become as welcome to them as flowers in spring.

True, at first, with strange ferrets you may be bitten; but it is not a very serious thing if you are, as ferrets' bites are never venomous, as the bites of rats often are. I have in my time been bitten by ferrets many dozens of times and have never suffered any ill effects. There, I think that is enough for your first lesson, so I will send it off at once and get it printed for you.

---

[9] Here, I presume that Barkley is referring to olive oil but it is possible that he is referring to some other vegetable-type oil.

2

THE first chapter of this lesson-book has gone to the printer, so I don't quite know what I said in it, but I think we had finished the home-life of the ferret and were just taking it out of its box. Different professors have different opinions as to what is next to be done with it. Many (and they are good men too) think you should put it into a box about eighteen inches long, ten inches high, and ten wide; the box to be divided into two compartments, with a lid to each, and with leather loops to these lids through which to thrust a pointed spade so as to carry it on your shoulder. I have tried this plan, but I have never quite liked it. I have found that after a heavy day's work the box was apt to get heavy and feel as if it were a grandfather's clock hanging on your back. Then the ratting spade was engaged instead of being free to mump[10] a rat on the head in a hurry, or point out a likely hole to the dogs. When a ferret was wanted, all the others would dash out and have to be hunted about to be re-caught. Now and then the lids came open and let all out; and now and then I let the box slip off the spade and fall to the ground, and then I felt sorry for the ferrets inside it!

No, I have always carried my ferrets in a good strong canvas bag, with a little clean straw at the bottom, and a leather strap and buckle stitched on to it with which to close it.[11] Don't tie the bag with a piece of string—it is sure to get lost; and don't have a stiff buckle on your strap that takes ten minutes to undo. Remember the life of a rat may depend upon your getting your

---

[10] Barkley's use of this word in the present context is certainly not related to the dictionary definition (to be silent or sullen, or, perhaps, to beg). I assume it to be a local slang usage meaning 'hit', 'clout' or 'wallop'.

[11] One wonders about ventilation! Most ferreters in Australia would, I think, use a box with a small wire window or, at least, some ventilation holes.

ferret out quickly. Never throw the bag of ferrets down; lay them
down gently. Don't leave the bag on the ground in a broiling sun
with some of the ferrets in it while you are using the others, or in
a cold draughty place on a cold day; find a snug corner for them,
if you can, and cover them up with a little straw or grass to keep
them warm.

If, when carrying your ferrets, they chatter in the bag, let
them; if is only singing, not fighting. I have never known a ferret
hurt another in a bag. Always bag your ferret as soon as you have
done with it; don't drag it about in your hand for half an hour,
and don't put it in your pocket, as it will make your coat smell.

When I have done work and turned towards home, I have
made it a rule always to put a dead rat into the bag, as I think it
amuses the ferrets and breaks the monotony of a long journey;
just as when I run down home I like taking a snack at Swindon
Station, just to divert my mind from the racketing of the train
and the thought of the hard seat. When you get home, give the
ferrets a rat for every two of them, if you can afford it, for then
they need only eat the best joints. If you have not many dead rats
and want to save money for the morrow, one rat for three ferrets
is enough for twenty-four hours; but don't forget to give them
water or milk.

I think I have said enough as to the management of ferrets,
and will go on to speak of the necessary tools. The chief thing
is a good ratting spade.[12] What the musket is to the soldier, the
spade is to the rat-catcher. You many get on without it, but you
won't do much killing. It should not be too heavy, but yet strong;
and, therefore, the handle should be made of a good piece of
ash, and the other parts of the best tempered steel, and the edge

_____

[12] This is quite a different tool to the ordinary spade or shovel which might be used in
Australia to dig out a warren or dig down on a 'dead end' (old breeding chamber). The
English version has a metal point on the handle for stabbing down to locate underground
passages, and a trowel-like spade-end. See Barkley's discussion of rabbiting spades in
chapter seven.

should be sharp enough to cut quickly through a thick root. The spike should be sharp, so as easily to enter the ground and feel for a lost hole.[13] This will constantly save a long dig and much time; besides, one can often bolt a rat by a few well-directed prods in a soft bank—not that I approve of this, as there may be more than one rat in the hole, and by prodding out one you are contented to leave others behind. No, I think the ferret should go down every hole challenged by the dogs, as then you are pretty sure of making a clean job of it.

Ratting spade.

Besides the spade, I have always kept a few trap boxes These are to catch a ferret should one lay up and have to be left behind.[14] I bait them with a piece of rat and place them at the mouth of the hole, and it is rare I don't find the ferret in it in the morning. I also take one of these traps with me if I am going where rats are very numerous; then, if a ferret stops too long in a hole, I stick the mouth of the trap over the hole and pack it round with earth and stop up all the bolt holes, and then go on working with the other ferrets. When the sluggard is at last tired of the hole, it walks into the trap, shoving up the wire swing door, which falls down behind it, and there it has to stop till you fetch it.

---

[13] Probing for a hole is not such an easy matter in most Australian conditions. Soils are often harder to penetrate and the actual burrow usually fairly deep.

[14] The term 'lay-up' refers to the habit of ferrets to sometimes curl up for a nap after having killed their prey and eaten part of it. In Australia, the term 'stick-up' is more common.

If I am going to ferret wheat stacks where rats have worked strong, I take with me half a dozen pieces of thin board about a foot long. I do so for this reason. The first thing rats do when they take possession of a stack is to make a good path, or run, all round it just under the eaves; and when disturbed by ferrets, they get into this run and keep running away round and round the stack without coming to the ground. Therefore, before putting in the ferrets, I take a ladder, and going round the eaves of the stack I stick the boards in so as to cut off these runs, and when a rat goes off for a gallop he comes to "no thoroughfare," and feeling sure the ferret is after him, he in desperation comes to the ground, and then the dogs can have a chance. I once killed twenty-eight rats out of a big stack in twenty minutes after the ferrets were put in, all thanks to these stop-boards; and though I ran the ferrets through and through the stack afterwards, I did not start another, and so I believe I had got the lot.

I think I have enumerated all the tools required for rat-catching. I need not mention a knife and a piece of string, as all honest men have them in their pocket always, even on Sundays.[15] Some rat-catchers take with them thick leather gloves to save their getting bitten by a rat or a ferret; but I despise such effeminate ways, and I consider he does not know his profession if he cannot catch either ferret or rat with his naked hands.

I must now turn to the subject of dogs—one far more important than either ferrets or tools, and one so large that if I went on writing and writing to the end of my days I should not get to the end of it, and so shall only make a few notes upon it as a slight guide to the student, leaving him to follow it up and work it out for himself; but in so doing I beg to say that his future success as a rat-catcher will depend on his mastering the subject.

But, before proceeding further, I am anxious to say a few words in parenthesis for the benefit of the Head Masters of our

---

[15] Not directly associated with anything specific to ferreting I imagine. Rather, just part of the essential impedimenta of an agricultural gentleman!

schools. Admirable as their academies are for turning out Greek and Latin scholars, I cannot help thinking a proper provision is seldom made in their establishments for acquiring a real working knowledge of the profession of a rat-catcher; and I wish to suggest that it would be as well to insist on all those students who wish to take up this subject keeping at school at least one good dog and a ferret, and that two afternoons a week should be set apart entirely for field practice, and that the cost of this should be jotted down at the end of each term in the little school account that is sent home to the students' parents. I know most high-spirited boys will object to this and call it a fresh tyranny, and ever after hate me for proposing it; but I do it under a deep sense of duty, being convinced that it is far better they should perfectly master the rudimentary knowledge of such an honest profession as that of rat-catcher, than that they should drift on through their school life with no definite future marked out, finally to become perhaps such scourges of society as MPs who make speeches when Parliament is not sitting. Judging from the columns of the newspapers, there must be many thousands who come to this most deplorable end; and if I can only turn one from such a vicious course, I shall feel I have benefitted mankind even more than by killing rats and other vermin.

Now I must return to the subject of dogs, and in doing so I will first begin on their master, for to make a good dog, a good master is also absolutely necessary. Anybody that has thought about it know that as is the master, so is the dog. A quiet man has a quiet dog, a quarrelsome man a quarrelsome dog, a bright quick man a bright quick dog, and a loafing idle ruffian a slinking slothful cur.

First of all, then, the dog's master must understand dog talk; for they do talk, and eloquently too, with their tongues, their ears, their eyes, their legs, their tail, and even with the hairs on their backs; and therefore don't be astonished if you find me saying in the following pages, "Pepper told me this," or "Wasp said so-and-so." Why, I was once told by a bull terrier that a country policeman was a thief, and, "acting on information received," I got the man locked up in prison for three months, and it just served him right. Having learnt dog language, use it to your dog in a reasonable way; talk to him as a friend, tell him the news of the day, of your hopes and fears, your likes and dislikes, but above all use talk always in the place of a whip. For instance, when breaking in a young dog not to kill a ferret, take hold of the dog with a short line, put the ferret on the ground in front of him, and when he makes a dash at it say, "What *are* you up to? War ferret! Why, I gave four and sixpence for that, you fool, and now you want to kill it! Look here (picking the ferret up and fondling it), this is one of my friends. Smell it (putting it near his nose). Different from a rat, eh? Rather sweet, ain't it? War Ferret, war ferret![16] Would you, you rascal? Ain't you ashamed of yourself? War ferret, war ferret!" Repeat this a few times for two or three days, and when you first begin working the dog and he is excitedly watching for a rat to bolt, just say "War ferret" to him, and he will be sure to understand. Should he, however, in his excitement make a dash at a ferret, shout at him to stop, and then, picking up the ferret, rub it over his face, all the time scolding him well for what he has done; but don't hit him, and probably he will never look at a ferret again.

In my opinion there is nothing like a thrashing to spoil a dog or a boy; reason with them and talk to them, and if they are worth keeping they will understand and obey. Mind, a dog must always obey, and obey at the first order. Always give and order in a decided voice as if you meant it, and never overlook the slightest disobedience. One short whistle should always be enough. If the dog does not obey, call him up and, repeating the whistle, scold him *with a scold in your voice.* Don't shout or bawl at him for all the country to hear and the rats too, but just make your *words*

---

[16] 'War' is presumably a contraction of 'beware'. Again, the usage may originate from the Old English wær – prudent, aware, alert, wary.

sting. If he repeats his offence, put a line and collar on him and lead him for half an hour, telling him all the time why you do so, and he will be so ashamed of himself that the chances are he will obey you ever after.

Put yourself in the dog's place. Fancy if, when you have "kicked a bit over the traces"[17] at school, the head-master, instead of thrashing you, made you walk up and down the playground or cricket-field with him for half an hour; but no, that would be too awful; it would border on brutality! But you would not forget it in a hurry.

We humans often behave well and do good, not because it is our duty so to do, but for what the world will say and for the praise we may get. Dogs are not in all things superior to humans, and in this matter of praise I fear they are even inferior to us. They most dearly love praise, and a good dog should always get it for any and every little service he renders to man. Remember, he is the only living thing that takes *pleasure* in working for man, and his sole reward is man's approbation. Give it him, then, and give it him hot and warm when he deserves it, and he will be willing to do anything for you and will spend his life worshipping you and working for you; for better, for worse, for richer, for poorer, he is yours, with no sneaking thoughts of a divorce court in the background.

There is another thing a master should always do for his dog himself and do it with reason. See to his comfort; see that he has good food and water and is comfortably lodged. Don't let him be tied up to a hateful kennel in a back yard, baked by the sun in summer and nearly frozen in winter; often without water, and with food thrown into a dish that is already half full of sour and dirty remains of yesterday's dinner. This is not reasonable and is cruel. When he is not with you, shut him up in a kennel, big or

---

[17] A term derived from the days of horse-drawn equipment. The 'traces' were the leather harness straps attaching to the implement being drawn. If the horse stepped over these, it was difficult for the driver to maintain control.

little, made as nearly as you can have it on the model of a kennel for hounds. Let it be cool and airy in summer and snug and warm in winter; keep all clean—kennel, food, dishes, water and beds. Don't forget that different dogs have different requirements; for instance, that a long thick coated dog will sleep with comfort out in the snow, while a short-coated one will shiver in a thick bed of straw. Picture yourself, as you tuck the warm blankets round you on a cold winter's night, what your thin-coated pointer is undergoing in a draughty kennel on a bare plank bed, chained up to a "misery trap" in the back yard, which is half full of drifted snow. Think of it, and get up and put the dog in a spare loose box in the stable for the night, and have a proper kennel made for him in the morning.

I once had a favourite dog named "Rough" that died of distemper. A small child asked me a few days afterwards if dogs when they died went to heaven, and I, not knowing better, answered, "Yes"; and the child said, "Won't Rough wag his old tail when he sees me come in?" When you "come in" I hope there will be all your departed dogs wagging their tails to meet you. It will depend upon how you have treated them here; but take my word for it, my friend, you will never be allowed to pass that door if the dogs bark and growl at you.

Don't suppose I am a sentimental "fat pug on a string" sort of man. Next to humans I like dogs best of all creatures. Why, I have made my living by their killing rats for me at twopence per rat and three pound farm, and I am grateful; but I like dogs in their proper place. For instance, as a rule, I dislike a dog in the house. The house was meant for man and should be kept for him. I think when a man goes indoors his dog should be shut up in the kennel and not be allowed to wander about doing mischief, eating trash, learning to loaf, and under no discipline. Now and then I do allow an old dog that has done a life's hard work to roam about as he likes, and even walk into my study (I mean kitchen) and sit before the fire and chat with me; but then, such dogs have established characters, and nothing can spoil them; besides, they are wise beasts with a vast experience, and I can learn a lot from them. It was from one of these I learnt all about the prigging policeman.

A young dog is never good for much who is allowed to run wild; everyone is his master and he obeys no one, and when he is taken out he is dull and stupid, thinking more of the kitchen scraps than of business. No, when I go to work, I like to let the dogs out myself, to see them dash about, dance around, jump up at me and bark with joy. I like to see the young ones topple each other over in sport, and the old ones gallop on ahead to the four crossways, and stand there watching to see which way I am going, and then, when I give them the direction with a wave of the hand, bolt off down the road with a wriggle of content. You might trust your life to dogs in such a joyful temper, for they would be sure to stand by you.

Thank you, young gentlemen; that is enough for this morning's lesson. You may now amuse yourselves with your Ovid or Euclid.

3

I AM a working man, or rather have been till I got the rheumatics, and as such I naturally stick to my own class and prefer associating with those of my own sort, and therefore I always keep working dogs.

I have often bred aristocratic dogs, dogs descended from great prize-winners and with long pedigrees, and among them I have had some good ones, honest and true; but as a rule I must say my experience proves that the shorter the pedigree the better the dog, and now if I could get them I should like to keep dogs that never had a father. Some people I know call me a cad, a clod, a chaw-bacon, etc., and they call my dogs curs and mongrels. Such men talk nonsense and should be kept specially to make speeches during the recess. I don't care to defend myself, but I must stand up for my dogs against all comers; and I assert boldly that, nine times out of ten, a dog with no pedigree is worth two with a long one. When I get a new dog I never ask who he is, or who is father was, but I go by his looks and his performances. There are dogs like men in all classes, who have either a mean, spiteful, vicious look, or a dull, heavy, dead one; such I avoid both in dog and man, for I find they are not worth knowing. Any other dogs will do for me, and even now, though I don't often go ratting, I have as good a lot as ever stood at a hole, and I don't think I can do better than describe them as a guide to students when they come to getting a kennel together.

First of all, I never give a lot of money for a dog—how can I with rats at twopence each?—but, if I can, I drop on a likely-looking young one about a year old who was going to be "put away" on account of the tax. I got the oldest I have now in the kennel in this way. It followed George Adams, the carrier, home one night, and to this day has never been claimed; and when the tax-collector spoke to him about it, he offered it to me, and I took it and gave it the name of "Come-by-Chance," but in the family and among friends she is now called "Chance."

If Chance is of any family I should think her mother was a setter and her father a bobtail sheep-dog;[18] but, then, I can't make out where she got her legs! She is red and white, with a perfect setter's head. She has the hind parts of a sheep-dog and evidently never had a tail; and her legs, which are very thick, would be short for a big terrier. Such are her looks, which certainly are not much to speak of; but if I had the pen of a Sir Walter Scott I could not do credit to the perfection of her character. For seven years she has been the support of my business, and I can safely say she has caused the death of more rats than all my other dogs put together. I say caused, for she is slow at killing and leaves this matter of detail to younger hands. If another dog is not near she will catch a rat and even kill it; but she has a soft mouth, and all the other dogs, except quite the youngest, know this, and, against the rule, will always dash in when she has a rat in her mouth and take it from her, and she gives it up without a struggle.

No, her forte is to *find* a rat. She is always in and out, up the bank, through the hedge, down the bank; not a tuft of grass escapes her, and she would hunt down each side of Regent Street and in and out of the carriages if she found herself there. She lives hunting. Nothing ever escapes her; one sniff at the deepest and most turn-about hole is enough. If the rat is not in, on she goes in a minute; but should it be ensconced deep down in the furthest corner, she stops and once and just turns her head round and says quietly to me, "Here's one." Then, whilst I am getting out a ferret, over the bank she goes, in and out the hedge in all directions, and never fails to find and mark every bolt-hole for the other dogs to stand at that belongs to the one where the rat is. As soon as I begin to put in the ferret, she will come over

---

[18] Another name for the Old English Sheep Dog.

the hedge, give herself a shake, and sit down and watch the proceedings, not offering to take a part herself, as she feels there are more able dogs ready, and that this is not her strong point. Suppose a rat bolts and is killed and the ferret comes out, Chance will never leave the hole till she has taken a sniff at it to make sure all the rats have been cleared out. I have never known her make a mistake. If she says there is a rat in, there is one without any doubt; if she says there is not, it is no good running a ferret through the hole. Should I be alone, with no one to look out for the ferret when it comes out on the other side of a bank, Chance without a word being said to her will get over and look out, and directly the ferret appears will come back to me and give a wriggle, looking in the direction of the ferret, and then I know I must get over and pick it up.

She has one peculiarity. When she followed George Adams home, seven years ago, she was shy and scared; but, as it was a cold night, George, being kind-hearted fellow, invited her to step indoors, an invitation she accepted in a frightened sort of way. On the hearth sat a little girl of three years old, eating her supper, and Chance, doubtless feeling very hungry, came and sat down in front of her and watched her with a wistful look. The child was not afraid and soon began feeding the dog, who took the pieces of food most gently from her fingers. When the child was taken up to bed, Chance secretly followed, and getting under the crib slept there all night. Only once since then has Chance failed to sleep in the same place, and that was the first night I had her. She was shut up in the kennel and never stopped barking all night. Since then she has always followed me home, eaten her supper at the kitchen door, and then gone off to her bed under the crib. Early in the morning she is again at my door and never goes near George's house till bed-time.

If Chance has no tail, the next dog on the list, "Tinker," makes up the average. He is a little black, hard-coated dog, with the head of a greyhound and tail of a foxhound. His head is nearly as long as his body, and his tail is just a little longer. In all ways he is proficient at rat-catching, except that he has been known to mark a hole where there was no rat; but his strong point is killing. He will stand well back from a hole, and it does not matter how many rats bolt, or how fast, each gets one snap and is dead and dropped without Tinker having moved a foot. I named him Tinker, for a tinker gave him to me "cos he wasn't no sort of waller."[19]

Then on my list next comes "Grindum," a mongrel bull-terrier, just the tenderest hearted, mildest dispositioned dog that ever killed a rat. He has but a poor nose and is not clever, but he has one strong point, which he developed for himself without being taught. It is this; when I am ferreting a thick hairy bank with a big ditch, Grindum always goes some ten-yards off and places himself in the ditch, and, let the excitement be what it will, he never moves; and should a rat in the thick grass escape the other dogs and bolt down the ditch, it is a miracle if it does not die when it reaches him. I have better and cleverer dogs, I know; but I think Grindum brings in as many twopences as any of them, and we are not going to part! The way I got Grindum is quite a little history, and I will tell it, though if you boys like, you can skip it and go on with a more serious part of your lesson.

Not far from where I lived there was, in a most out-of-the-way corner on a common, an old sand-pit, and in this a miserable dilapidated cottage, consisting of two rooms. This for some years had been empty, but one fine morning was discovered to be inhabited by a man, his wife and two children—a boy of twelve and a girl of seven—and a bull-terrier. No one knew anything about them or where they had come from, and when the landlord of the hut went to eject them, he found them in such a miserable half-starved condition that he left them alone.

Our parson called on them three times—the first time the wife

---

[19] Perhaps meaning "because he was of no value" or "he was no good"?

told him they did not like strangers and parsons in particular; the second time the husband told him to clear out sharp, or he would do him a mischief; and the third time the man took up a knife and began sharpening it, preparatory, he said, to cutting the parson's throat!

Two months after this the man, after sitting drinking in the village pot-house all the morning, stepped round to an old mid-wife and asked her "to come and lay his wife out." The woman went and did her work and said nothing at the time, but later on it was whispered about that she had told some of her pals that "the poor critter was black and blue, and it was on her mind that the husband had murdered her!" After this, as I passed the cottage, I often saw the two children sitting on a log of wood outside with the bull-dog sitting between them. None of the three ever moved out; all blinked their eyes at me as I passed, as if they were unaccustomed to the sight of a fellow-creature.

Two or three months passed, during which the man was constantly drinking at the village public-house; but he always left at sundown—"to look after the kids," he said. Then there was a poaching fray on a nobleman's estate near. Six keepers came on five poachers one moonlight night. There was a hard fight, and at last the keepers took two of the men and the other three bolted, but one was recognized as the man from the sand-pit and was "wanted" by the police.

A few nights after this I was walking down a lane in the dark near my house, when the sand-pit man stepped out of the hedge, leading his dog by a cord, and turning to me said, "Here, master if you want a good dog, here is one for you; I am off to give myself up to the police, and I am going to turn Queen's evidence against my pals." I replied that I did not want such a dog, so he said, "All right, then I'll cut his throat," and then and there prepared to do so. This was more than I could stand, so I took the cord and led the dog away, but before doing so, I asked, "How about your children?" He gave a short laugh, and said, "They would be properly provided for." If afterwards turned out that soon after leaving me he walked straight into the arms of two policemen, who saved him the trouble of giving himself up by taking him into custody.

I led my new dog home and tied him up in the corner of an open wood-shed, giving him a bundle of straw and a dish of bones, and by the starved look of him I should say this was the biggest meal he had ever had in his life.

I sat up late that night reading, and all the time in a remote corner of my mind the sand-pit man, the two children and the dog kept turning about, till at last, about midnight or later, I thought I would go to bed; but before doing so I made up my mind that I would see my new dog was all right. I lit a lantern and stepped out of the door and found it was blowing and snowing and biting cold. Mercifully I persevered and reached the wood-shed, and what I saw there by the light of my lantern did startle me. There was the bull-dog sure enough laying curled up in the straw blinking hard at me, but—could I believe my eyes?—there lying with him, with their arms entwined round each other and round the dog, were the two children from the sand-pit fast asleep, but looking so pale and pinched I thought they must be dead.

I will give place to no man living at rat-catching and minding dogs, but here was a pretty mess for I am no good with little children; so putting down my lantern, I hurried back to the house and got two rugs and with them wrapped the children and dog up snugly. Then I went in and woke up my wife, who had already gone to bed, and called some other women who were in the house, and after telling them what I had found, I made up a big fire in the kitchen and put on some water to boil. In a very few minutes my wife was downstairs and battling her way with me off to the woodshed. I untied the dog and moved him away from the children. This woke them both, and they sat up and rubbed their eyes, and the poor boy appeared almost scared to death, but the little girl was quite quiet, and only watched his face with a sad careworn old look which I pray I may never see on a child's face again.

My wife is really smart with little children, and in half no time she was on her knees crooning over them, and soon she had the girl in her arms; but when I attempted to pick up the boy he only screamed and struggled, and kept calling out, "Grindum, Grindum! I won't leave Grindum. I shall be killed if I leave Grindum. Let me stay with Grindum." I assured him he should not be separated from Grindum "never no more," and at last I partially quieted him, and he allowed me to carry him into the kitchen and place him on a stool in front of the fire with his sister, while his beloved Grindum sat by his side blinking as if nothing unusual had taken place, and as if he had done the same each night for the last three months and felt a little bored by it.

The first thing to be done, my wife said, was to feed the children, and while she and the other woman busied about getting it ready, I sat and watched them. Both were remarkably pretty; both dark, with finely cut features, big eyes and thick soft black hair; but yet in different ways both had something sad about them. The boy never sat still for a moment, but kept glancing fearfully at me, then at the women, and then and the door, as if he expected something dreadful to happen, and all the time kept grasping the arm of his little sister with one hand as if for protection, and clinging to the soft skin of Grindum's neck with the other. If he caught my eye, or if I spoke to him, he flinched as if I had struck him, and turned livid and tugged so hard at Grindum's skin that the poor dog's eyes were pulled into mere slits, through which I could see he yet went on blinking at the fire. The girl sat half turned round to the boy and never took her eyes off his face, looking the very essence of womanly pity and love. Now and then when he suffered from paroxysm of fear, she would softly stroke his face, which appeared to soothe him instantly; but nothing she could do could ever stop the wild restless look in his eyes or prevent his glancing about as if watching for some dreadful apparition. It was sad, sad picture, made doubly striking by the utter stolidity and indifference of that awful dog Grindum.

Soon hot basins of bread and milk were prepared, which both children ate ravenously, then they were put into steaming hot baths, washed, dried, combed, and wrapped in blankets; but when we attempted to take them up to the nice warm beds that had been prepared for them, there was the same wild terrified cry from the boy from Grindum; and to pacify him the dog had to be taken upstairs with them, and half an hour later, when my wife and I peeped into the room, we saw the two children locked in each other's arms fast asleep, with Grindum curled up on the bed next to the boy, yet blinking horribly, but perfectly composed and making himself at home.

How those two children found their way that night through a blinding snow-storm to their only living friend, the dear blinking Grindum, I never could find out. All I could ever get from the boy was, "Oh I always go where Grindum goes!" and the little girl could only say, "Jack took me." My wife says angels guided them. Maybe she's right, but I hardly think angels would be likely to go about on such a night; still my wife went out in the snow and wind to the shed and got out of her snug bed to do it, but then she put on a pea jacket and clogs, and that makes a difference.

This is a tiring long story to write, and I have not quite done it yet, for I must finish with the sand-pit man. He was tried, convicted and got three years. A year after he had been in prison he tried to escape by getting over a high wall, but in doing so he fell from the top and broke his back. He lingered some days and seemed to find a pleasure in telling the prison parson of all his misdeeds and in boasting of them. There was a long list, but only the last part of his story will serve for "the use of schools." It appears from what he said that, after he had given me the dog, he had intended to steal back to his house and take the two children to a deep pond and there drown them. Then, free from family ties, he hoped to get away and ship himself off to America. He also said that in a fit of rage he had thrashed his wife to death with his fists, and that his boy from having seen him do it had gone mad with fear, and was so bad, especially at night, that if he had not got a bulldog sleeping with him as a sort of a friend, he would go into a fit with fear and was often unconscious for hours.

It was an ugly story, and I am glad to say with the death of the sand-pit man the miserable part of the children's life ended. The girl is now twelve years old and has never left us. She is as sharp as a needle and honest as old Chance and as good. She is having a good education, thanks to our Rector's wife, and could if need be earn her own livelihood, but we are not going ever to part with her.

The boy Jack was a great trouble to us at first. For months he would not be parted for a moment, day or night, from Grindum, and the dog actually had to go to school with him; but the master utterly failed to teach the boy as far as ABC in his alphabet, and the dog not to blink; and so, one fine day, I had both returned on my hands as hopeless ignoramuses, I could not keep a blinking dog at home with idleness, so I took him with me ratting, and as Jack would not be parted from the dog, he had to come too. Everyone says the boy is "cracked." He is queer, I will allow, but if you will find me a better hand at rat-catching in all its branches, I should like to look at him; and besides, if Jack is cracked, then I like cracked boys, for I never came across one more obedient, more truthful, or more steady, and I find him a perfect treasure on the other side of the bank at the bolt holes.

Jack never mentions the past, and I should be inclined to think he had forgotten it, only if he is parted from Grindum for a short time he becomes wild looking about the eyes again and restless. At such time his sister, who mothers him much, will sit by him and stroke his face softly, when he will quickly recover himself.

I don't know what will happen when Grindum "blinks his last," but the boy begins to follow me about and seems to cling to me, and by that time I hope I shall be so well liked by him that I may take Grindum's place.

Just two words more about Grindum and I have done. One is that he first time Grindum caught a rat, he picked it up by its hind leg, and the rat made its teeth meet through his nose. He softly put the rat down and it escaped, and I made my sides ache and greatly astonished all the other dogs laughing at this great soft beast as he sat on his haunches licking the blood as it trickled from his nose, and staring up into the sky with a far-off vacant look, blinking worse than ever.

The other word is this. Though Grindum is a bull-dog with an awful "Crush your bones, tear your flesh" look, he is just the gentlest-hearted beast out, and there is not a puppy in the kennel, nor a child in the village, who does not know this and impose on him shamefully. Only last Sunday I had to stop a small child of five from driving off a four-wheeled car, using Grindum as a horse. Once, and once only, Grindum showed his temper. A big lout in the village threw a stone at him. Grindum only blinked, but Jack saw it and hit the lout, who being twice Jack's size turned upon him and knocked him down. In half a minute Grindum's teeth had met three times in the lout's calves and his trousers required reseating, and in three-quarters of a minute Grindum was sitting with a bland expression of countenance, blinking with both eyes at the sky.

Now to continue my lesson on ratting dogs. I have two others, Pepper and Wasp—one a badly bred spaniel, and the other a terrier of doubtful parentage. They are both nice cheerful young dogs that it is a pleasure to see either at play or work, but they are yet young and too apt to get excited and wild. They will, when a rat is out of his hole, in a hedge, dash up and down the entire length of the field, making enormous jumps in the air, during which time they listen keenly for the rustle of the rat in the grass; and once, but only once, Pepper gave a yap when so rushing about, but I spoke so severely about this disgustingly low habit that he has never done it again.

Wasp is specially good at water, and I have taught her to come to me directly a rat is bolted with a plunge into a pond, and I carry her high up in my arms round the pond, and when the rat approaches the side, Wasp from her high vantage ground will dive down upon it and have it in an instant. Both dogs are quick killers and will, I am sure, in time be perfect; but as yet I do not think myself justified in putting them into a higher class with such dogs as Chance and Tinker.

There! That is all for to-day, young gentlemen. Resume your Cicero, and, while you are preparing it, I will go to my room and look over the impositions I set you yesterday. It is understood that for "look over impositions" we may read, "Smoke cavendish in a short black pipe."

4

WHAT do you say, boys? Shall we drop this and have a day's outdoor practice? To tell the truth, I don't think much of this book-learning, especially if the book is written by myself; but I do believe in practice. Come along! It is in the middle of October—just the nicest time of the year and the very best for ratting, for the vermin are yet out in the hedges, fine and strong from feeding in the corn, and with few young ones about. Come, Jack, we'll get the ferrets first; and off I go with the boy to the hutch, while the dogs in the kennel, having heard our steps and perfectly understanding what is up, bark and yap at the door, jump over each other tumble and topple about like mad fiends. Before I get to the box I hear the ferrets jumping up at the sides, and when I open the lid half a dozen are out in a moment, and these I bag as a reward for their activity. I throw the others a rat to console them for being left at home, and, giving the ferrets to Jack, I strap on a big game bag, take up my spade, return and let the dogs out, and off we start.

Step out quick, Jack; there are three miles to go before we get to work, and it is 8 a.m. and I expect a big day. Yes, Chance, old lady, a fine day—a perfect day—a day to make both the feet and the heart light and every human sense rejoice. There has been just a little frost in the night; you can see that by the way the elms have spread a golden carpet under their branches in the lane and by their leaves that yet keep falling slowly one by one in the fresh, but dead still, air, and by the smell of the turnips, the fresh stubble and the newly turned earth behind yonder plough. The sun shines, cobwebs are floating through the air and get twisted round one's head, and far and near sounds such as a cart on the high road, a sheep dog barking, a boy singing, birds chirping, insects humming, the patter of our own feet, and the whispering of the brook under the bridge, all form part of a chorus heaven-sent to gladden the heart of man. I have heard tell, Chance, or I have seen it in a book, or I have felt it myself, I don't quite know which, that those who in youth have had such a walk as this, and have heard the music, smelt the perfumes and seen the sights (that is if they were blessed with eyes to see, ears to hear, and hearts to take in), have never forgotten it. The memory appears for a time to pass away amidst the struggles of life, but it is never dead; to the soldier in battle, to the statesman in council, or the priest in prayers, to those in sorrow or in joy or in sickness, there may come, no one knows from where, no one knows why, a golden memory of such days, of such a walk. Perhaps it is only a gleam resting but a second upon the mind, and perhaps leaving it saddened with a longing for days that are past, but yet I think making one feel a better man, giving one courage and hope, reminding one that, hard as the battle of life may be to fight, dark and gloomy as the days may be just now, another morning may arise for us, far, far more bright and glorious and joyful, one that will not be shadowed over by a returning night; but then that is only for the brave, the honest, the truthful—for those who are up early and strive late, never beaten, never doubting, always pressing forward.

But, come out of that, Wasp! Don't you know that cows kick if you sniff at their heels? Tinker, old man, keep your spirits up; Pepper, come back from that wood, for it is preserved. Yes, Jack, I think I'll fill my pipe again. Baccy does taste good on a day like this; but what doesn't? I feel like a ten-year-old and as fit as a fiddler. Grindum, give over blinking and don't look so benevolent. No, Chance, no, old lady, I can't pull your tail, for you haven't got one. What, Jack, you say I haven't spoken for the past mile? Well, I suppose I have been thinking, and my thoughts have not been wholly sad ones. Open the gate; here we are; and you get over on the other side of the hedge and don't talk or make a noise, for I can see by the work the rats s-w-a-r-m.

Steady, dogs, steady! And so we start.

The hedge is just what it should be, and if it had been made for ratting it could not be better. A round bank of soft earth, a shallow ditch with grass, little bush or bramble, and a gap every few yards. There is a getaway in the middle, which will make a hot corner later on when Grindum has taken his stand there; and there is a pipe under the gateway, the far end of which I shall close. The rats have never been disturbed, for the runs are as fresh as Oxford Street, and I have already seen one or two rats run into the hedge lower down from out the wheat stubble, and, there! that whistle has sent a lot more in. Steady, Wasp! Well done, Chance; you have marked one in that hole near you, or more than one, is there? Well, the more the merrier! Stand, dogs, stand! Are you ready, Jack? And in goes a ferret as lively as quicksilver and as fierce as a tiger.

For a minute all is quiet; then a slight stir on the other side and two snaps of Tinker's lantern-jaws, and two rats dead; three others out of a side hole are killed by Wasp, and three others accounted for by Grindum, and that fool Pepper is racing and jumping down the hedge a mile off. Whistle! whistle! And back he comes, and at that moment Jack picks up a ferret on the other side, it having gone through the hole. Chance sniffs at it and says it is swept clear, and I block it up with my heel, and Jack does the same to the bolt-hole, so that if a rat does come back later on the dogs will have a chance; and then on we go a few yards to the next hole which Chance marks. This time the ferret went in like a lion and came out like a lamb, with the blood running out of the side of it's face; and whilst I am examining the bite, a real patriarch rat bolted at a side hole near Pepper, who strikes at it, misses taking a proper hold and gets it too far back, and the next moment the blood is pouring from a bite about his eye; but the rat is dead, and Pepper but little the worse.

I thought it was too late in the year for young ones, but it was not, for at the next hole we came to the ferret got into a nest, killed a lot of young ones and "laid up," and, as I had not a box-trap with me, I had to dig it out. This took some time, as I lost the hole, and Jack, whilst down grubbing with his hands, broke into a wrong one in which the old rat was ready for him, and at once bit him through the end of his finger. Jack sucked it well and did not mind, but I did not much like the appearance of things, for in half-an-hour I had had a ferret laid up, and a dog and a boy bitten badly by rats, and these bites are often poisonous. Fortunately this time Jack took no harm and was soon well. As soon a Jack pulled his hand out of the rat's hole, Pincher put his long nose in, and all was over in a minute. Soon after I came on the ferret curled up in a nest of young rats, all minus their heads; and so that ferret, from being gorged with food, was no more good for work, and had to be put away with the bitten one.

After this we got on much faster; the holes were close together, and even with the greatest care lots of rats bolted and went forwards, but I would not allow the dogs to disturb fresh ground by following them. Some went back, and Pepper and Wasp had a good time, for I let them follow and work them alone, having stopped all back holes after ferreting them. Now and then, Jack and I had to go back, as there was an old pollard tree[20] covered with ivy, and many of the rats got up that, and Pincher had to be lifted up into the crown to displace them, and then when they jumped down, three or four at a time, there was a grand scrimmage.

When we had got twenty yards or so from the gateway, Grindum went forward and stood there and killed a dozen rats that tried to pass, and a lot more went into the pipe under the roadway. These we left alone, only after we had passed we stopped up the open end and opened the shut one, so that in future rats going back might wait quietly in the arch till we were ready for them. By the time we had got as far as the gate it was

---

[20] This does not refer to a species of tree, but to the form of its growth. Pollard trees are trees whose head limbs have been pruned back.

just noon, so we called the dogs back to a tree we had passed, and then Jack and I sat down and paid attention to the game bag, which was well provided with cold meat and bread and cheese and a bottle of beer.

I am not a good hand at picnics and never was. I mean those big gatherings, with ladies, lobster salad, hot dishes, plates, knives, spoons, champagne, etc. I find the round world was created a little too low down to sit upon with comfort; my knees don't make a good table; flies get into my beer and hopping things into my plate. I have to get up and hand eatables about; things bite me, and more creep about me, and it does not look well to scratch. The hostess looks anxiously about her glass and plate; someone has forgotten the salt, and someone else the corkscrew. The host, be he ever so sad, makes fun, and made fun is magnified misery to me. No, I don't like picnics; I would rather be at home and feed upon a table; and yet a snack at noon-day, after hard work, sitting under a tree, with your hands as plates, with a good "shut-knife," a silent companion and the dogs all round you is pleasant. Double Gloucester then equals Stilton, and bottled beer nectar; and then the pipe in quiet, while Jack takes the dogs, after they have finished the scraps, to the pond to drink. Talk of Havanas! Well, talk of them, but give me that pipe as I loll, half asleep, resting against the tree, my legs spread out, and my hat tipped over my nose. I half close my eyes and go nearly to sleep, but keep pulling at the pipe, and half unconsciously hear the leaves whispering about, the insects humming, the stubble rustling, the trembling of a thrashing machine, and the rush of a train in the far distance. Jack returns from the pond, throws himself on the ground on his face, kicks his legs in the air and whistles softly, with the gentle Grindum blinking beside him. Chance and Tinker lie full length on their sides and go to sleep. Wasp stretches on the ground, with her legs out behind her, and drags herself about with her front feet. Pepper sits down, scratches his ear, and then dashes at a passing bumble bee, and all becomes a pleasant jumble of sights and sounds; but, with a start, I recover myself, drop my pipe, topple my hat off and lose my temper, for that everlastingly restless, volatile, good-for-nothing, ramshackly beast, Pepper, has been and licked me all up the side of the face! The dream, the quiet, the rest is all broken, so, jumping up, I tip my pipe out on the heel of my boot, give a stretch, grasp the spade, and off we go to finish our job.

For three hours we work our way on, and a line of dead rats on the headland marks our progress, till at last we reach the bottom of the field and our bank is done. Pepper has got three more bites, another ferret is done for by a nip on the nose, and Jack has torn his trousers and is very dirty; but there is yet the drain pipe under the gate to attend to, and it is getting on in the day. I cut three or four long sticks and tie them tightly together, and then to the end of this fasten a good hard bunch of grass, and back we go to the drain. I go to one end with Grindum and Pincher, whilst Jack takes the sticks, Pepper and Wasp to the other end, and gently and slowly shoves the sticks through. Two venturesome rats bolt at my end and are killed. When the sticks appear I grasp them and gradually draw the whisp of grass into the drain. It fits tight and takes some pulling, but it comes steadily along, wiping all before it. Faster and faster the rats bolt and are killed, and even old Chance, who began by watching us, gets excited and joins the sport. Pepper and Wasp dash in for a last worry, which is over in a few minutes, when twenty-four rats are cast by Jack up on the bank. Well done, dogs! Well done, good dogs! woo-hoop, woo-hoop! Good dogs! That's the way, my boys! woo-hoop! woo-hoop! And the dogs roll on the ground, stretch, wipe the dirt out of their eyes with their paws, and rub their faces in the grass.

Jack goes backwards and forwards and collects the spoil, and we count up seventy-three real beauties, a few of which I really think should be fourpenny beasts, they are so big. Never mind, seventy-three rats at twopence each comes to twelve and twopence—not such a bad day's work; and, Jack, you shall have a hot supper to-night; and oh, you dogs, you dogs, think of the supper I will give you! Bones with lots of meat on, oatmeal and such soup! Think of it, dogs! Think of it! And so the work ends, and all are happy and contented.

Three miles down turning twisting lanes to reach home, Grindum and I first, then Jack, and the rear brought up by the long and now a little drooping tail of Tinker. All have had enough; even the volatile young Pepper trots slowly, and therefore looks ever so much more business-like.

Before we start the shades are falling, and as we trudge along nature's evening vespers speak of the closing day. Workmen sitting sideways on quiet harnessed cart-horses stump past with a friendly "Good night, neighbour, good night!" Women with children in "go-carts" bustle past in a hurry to get home and fetch up the supper. Farm horses are drinking in the pond or browsing on the rank grass at the side; sparrows are chattering in the old alder bush before going to bed in the ivy on the church; pigs in the homestead are calling for their supper; the cows pass us coming home to be milked; rooks fly steadily to the old elm trees near the Manor; and a robin pipes clear and shrill on the roof of the shed in the cottage garden. There are partridges calling out "cheap wheat" in the stubble, and pewits crying on the meadows. Cock pheasants noisily flutter up to roost in the firs, and the old doctor standing at his door makes soft music with his violin.

The parson joins us and has a cheery word for all, especially the dogs, who are all his personal friends; and so we jog on and reach the village, where the wood smoke rises straight in a blue cloud from the cottage chimneys, and the fire light sends a ruddy gleam across the roads. Groups of men and boys stand about resting, little children race and play, and oh, such a delicious whiff of something stewing, with a little bit of onion in it, comes from the open door of the village ale-house! And this reminds us all that our suppers are near, and we finish the evening's walk quite briskly.

No need to say, "Kennel, dogs, kennel!" All go in of their own accord, and in five minutes are busy at their savoury-smelling *hot* supper. The ferrets are fed and locked up, and then, unlacing our boots at the back door and kicking them off, the day is done. Supper, rest and quiet, a pipe, a book, bed and happy dreams are all before us.

"Now, Croker, minor, you will go to the Doctor's study before school to-morrow. You have been most inattentive, and it is not the first time I have had occasion to speak to you. You can go now, but don't forget that this is tub night, as you all have done on the last four occasions. If I have further complaints on this head from the matron, I shall take you all out for a long day's rat-catching, so I advise you all to be very careful." Five minutes later this master is smoking in his room and says to another master who is doing the same, "I say, Potts, do you know I think these new lessons on rat-catching are all very well, but I think are beyond the capacity of school children. Why, they strain *my* mind, and I think they should only be taken up at the universities and during the last term; and then the boys and girls do so hate them," etc.

5

"CROKER, minor, have you been up to the head-master? Yes? Then sit still and don't fidget. Boys and girls, pick up your books on rat-catching, and we will resume yesterday's task."

The last chapter treats of a prime day's rat-catching, where rats were numerous and known to be numerous; but don't suppose all days are like this, for if you do you will be sadly disappointed, and you will have a lot to learn, for there are days, and very pleasant days too, when you will have to walk mile after mile to find a rat, and even then not be successful; but you will be out of doors in the fresh air, with devoted companions and something fresh to see at every step, if you keep your eyes open. Don't get disheartened, and above all things never say, "Oh, it is no good looking here or looking there for a rat; there is sure not to be one. Come on and don't waste time." You often find them in the most unexpected places.

I once went three times to the house of an old lady, being sent for because there was a rat that came each night and took her hen's eggs and carried off young ducks and chickens. I spent hours looking for it in hedges, ditches, sheds, out-houses and stable, and even put Tinker up on the roof of all the buildings, thinking the assassin might be under the tiles; but it was no go.

Night after night the plunderer came, and I began to see that the old lady did not think much of me. At last, one afternoon, I called again and began operations by asking to have a dog that was tied up to a kennel in a back yard led away, as his barking disturbed my dogs. This was done, and a minute afterwards Chance was sidling round the kennel, staking her reputation upon the rat being under it. I got out a ferret and looked round the kennel, and was utterly disgusted to find it was placed firmly on hard ground without a vestige of a hole. I am sorry to say I went so far as to sneer at Chance and tell her she did not know the difference between a dog and a rat. She herself for moment seemed in doubt, but the next she went *inside* the kennel and stood at a hole in the plank floor. I put the ferret back in the bag and, taking hold of the kennel, tilted it up, and in an instant the dogs had a vicious-looking old monster dead.

Now the only possible way that rat could have got in and out of his house was by passing the dog as he slept, and yet the old lady and her gardener assured me that the dog was as keen as mustard after rats.

I once killed a rat inside a church. I found it during a long sermon, but for the life of me I can't remember what that sermon was about. I was sitting in a seat opposite about a score of village school children, and suddenly I was struck by their appearance, and the thought passed through my mind, "How like humans are to dogs! Why, those children look just like my dogs when they find a rat, especially that flaxen-haired girl with a front tooth out." Then I noticed that they were all looking in one direction, and so I looked there too and saw a rat sitting with just its nose out of a hole which ran under the brick floor, apparently listening to the sermon. The next morning the parson and I went to the church. I took one ferret and only Tinker. I chose Tinker because he was black and rather clerical looking. The rat was at home, and we had it in five minutes. This was one of the few times I ever did rat-catching with my hat off, and it felt very queer.

Again, I once killed a mother rat and a lot of young ones which I found in the stuffing of a spring sofa in a spare bedroom at an old manor-house. There were rats in the walls, and "Mary Ann" had often seen a rat in the room when she went in to dust, and it had given her "such a turn." This time I took all the dogs with me, and we were followed by the lady of the house, four dreadfully pretty daughters and "Mary Ann." Madam and Mary Ann got on the sofa, standing, and the four daughters stood on four chairs round the room. All six clasped their clothes right round their ankles—why, I never could think. This was the only time in her life that I ever found Chance a fool. Directly she got into the room, she wriggled and twisted, turned her head this way and that, threw herself on her back and fairly grovelled. Wasp, Pepper, and the long-tailed Tinker were nearly as bad, and it was plain to see they were shy and bashful in such a gorgeous room and surrounded by such a galaxy of beauty. It was the soft-hearted Grindum who saved us; he blinked much, but directly I said, "Hie round, dogs! Hunt him up! Search him out!" he went to work—up on the bed, round the room, behind the furniture, and at last began sniffing round the sofa. I got hot all over, for I thought he was mistaking an aristocratic lady and her hand-maid for rats; but no at last he went under the sofa, and turning over on his back began to scratch at the underside of it up above him. Madam and Mary Ann jumped off, and the latter felt another "turn"; then both took refuge on chairs and again clasped their clothes tight round them. I turned the sofa up on its back, and there through the sacking near a leg I found a nice round hole into the interior among the springs. I put a ferret in, and in a minute there was a rush and scuffle, the sofa seemed alive, and then three or four small rats bolted out and were accounted for; another squeak and rush, and out came the mother and was quickly dispatched; then, as the ferret did not come out, I ripped the sacking and found it eating a deliciously tender young rat. I bagged the ferret, and while I did so, Grindum killed three or four small ones. I afterwards found that the rats had eaten through the wainscot and so got into the room. The rest of the afternoon was spent in turning over all sorts of furniture, including beds, and hunting through each room with the dogs; but we found no more rats as inside lodgers.

Three or four months after this episode, rats swarmed in the walls of the same house and behind the wainscoting, and my professional services were called in to get rid of them. How they got into the house I never discovered, for there were no holes from the outside, and no creepers on the walls for them to mount by and get on the roof; the drains did not appear to communicate with the inside of the house, and all the door fitted tight. Equally puzzling was it, now that they were inside, to get them out, for I dare not put ferrets in, for fear they should kill a rat and leave it to decay and smell for months.

I tried various plans. I got a live rat, tied a ferret's bell on it, and turned it loose, and for days after it was constantly heard tinkling inside the walls; but it did not drive the rats away. I singed the coat of a rat, put tar on the feet of another and turned them loose; but it was no good. At last I took possession of a wood-house in a cellar down the basement, from which a short passage led to other cellars, and in the walls of these there were many open holes. First of all I went carefully over the wood cellar and made sure there were not holes in it; and then, putting in a few faggots to give shelter to any nervous young rat, I started each night to feed them with delicious balls of barley-meal, which were made up with scraps. In this way I gave a rats' supper-party each night for three weeks, and each morning I found clean-swept dishes. At last the fatal day arrived. A string was tied to the handle of the door leading up into the kitchen, the food was placed in the dishes as usual about ten p.m., and all the household, except myself, went to bed. I sat over the kitchen fire reading my paper till distant clock struck midnight, and then I gave a sharp pull to the string and heard the door bang to and the fastening fall, and I knew I had them. I lit a big glass lantern, went round to the stables and let out the dogs, took them to the cellar window and slipt them through quickly, squeezing myself through after them and shutting the window again. In half no time fifty rats were killed, and all the dogs, except Tinker, pretty badly bitten; but they were used to that and did not care. Then I locked the back door behind me, taking the key home to bring back in the morning when I called to be paid eight and fourpence for my night's work. Three times in the next three months I went through a similar performance, and the first time I killed twenty-eight rats, the second seven, and the third time only two, and these were old bachelors. Then every hole in the walls was filled up with a cement made up with broken glass, and I have never heard of a rat in that house since. Before I forget it, let me tell you that if a rat dies in the wall, or under the floor of a house where it can't be got at, its whereabouts can be discovered in this way, provided the weather is warm. Take a butterfly net over to the butcher's shop, and there catch a dozen bluebottle flies, and taking care not to hurt them, slip them into a glass jar and tie a rag over it. Return to the room where the smell is, and, shutting the door after you, let your pack of flies loose and sit down to watch them, and in half-an-hour you will find they are all buzzing round one spot. Have this spot opened out, be it wall or floor, and there the dead rat will be found.

Has the bell rung? Yes, half a minute! Put your books away, form two and two outside, and I will take you for our usual walk. Croker minor, the top part of Jones' leg was not made to stick pins into. If I see you do it again, I shall give you a rat to catch, so be careful!

6

I TRUST that, in the five chapters I have written, I have said enough to give some of my scholars a slight taste and liking for the profession I am advocating, and in some small degree have weaned their young affections from such pernicious pastimes as studying classical authors, doing sums, and cutting their names on their desks. If I have not done this I have written to little purpose, and I fear the next chapter will damp off a few who have only followed me and my dogs on fine days in pleasant paths; but I may as well tell you at once that life is no more all beer and skittles in rat-catching than it is in such minor professions as the Army, the Church, the Bar, school-keeping, etc.; and just to see if you are "real grit," boys and girls, I will show you another picture.

Jack, get the ferrets while I let the dogs out. We *must* go and see if we can find a few rats, for it is a week since the ferrets had flesh, and we shall have them getting ill; and, Jack, bring four in the little bag, and put that inside your game-bag, for it looks like rain, and I don't like to see them half-drowned. Yes, it does look like rain, though as yet it is only a dull, misty, chilly, day in mid-November down here in the country, but in London it is a thick black fog, and all work is being done by gaslight. It is bad and depressing here, but ever so much worse there; so cheer up, dogs, and step out, Jack. We will go down by the beck and home by the clay-pits, for I know of no other place near where we are so likely to find a few rats, and I don't want to make a long day of it.

Go over the bridge, Jack. You take that side with Chance and a young one, and I will do this side with the other dogs. Hie in, dogs! Search him out, lads! And on we go, but in two miles we only kill a waterhen that Pepper catches as it rises out of some sedges, and which goes into my bag to replenish the ferrets' larder. The mist hangs low, the bushes are wet, the ground soft, and there is a dreary sigh in the wind. The cattle are eating fast, as they always do before rain; and the sheep, startled by the sight of the dogs, caper and jump as they gallop all down the meadow; and again their playfulness warns me of a wet tramp home. Some young colts stand at the door of an open shed, dull and depressed looking, and the horses ploughing on the sides of the hill send up a thick steam. No birds twitter or sing, no insects hum, distant sounds are muffled and indistinct. The teams in the wagons on the road hard by creep along and take little notice beyond a toss of the head at the carter's whip as he walks beside them with a heavy step cracking it. The only brisk thing to be seen is the doctor's gig as it whisks past.

"Hie up, dogs! Shake yourselves and don't go to sleep! Come over, Jack; I have had enough of this brook; and if we don't find at the clay-pits, home we go." And we trudge off to some ponds half a mile further away. They are well-known to both men and dogs, and the latter bolt on ahead and arrive first; and when we come up we find them all clustered round a hole in a high bank 'midst thick dripping bushes. In goes a ferret, but not in the way I like to see. There is a no hurry, no ecstatic wriggle of the tail as it slowly draws itself into the hole. Then all stand round expecting to see a rat take a header into the pond; but no, five minutes pass, and Pepper begins to move, and is told to "stand." Ten minutes pass, and Jack gets restless. Fifteen minutes, and I begin to shift my feet, which are planted deep in sticky mud by the side of the pond, and just then the first drops of rain appear. Ah, there is the ferret! Jump up and get it, Jack. But before he can do so, it has drawn itself into the hole backwards, which means that it has killed a rat inside and that it only came out to tell us so, and that it was going back to have a good long sound sleep curled up by the rat's warm body. There is nothing for it but to dig it out; and oh, what a dig, all among roots and thorns on the sloping sides of the pond, in thick sticky clay, with the rain coming down in a steady pour! Jack hunches his back and leans against a tree, Pepper and Wasp wander away down a ditch and scratch for an hour at a drain that has a rabbit in it, and the old dogs sit and watch me and drip and shiver. I dig here, I dig there; I slip and fall on the bank; the water mixed with yellow clay runs up my arm from the spade, and yet that beastly ferret sleeps peacefully in its warm bed. I lose the hole, come down on roots as thick as my leg and stones that strike fire as the spade strikes them; and so two hours of discomfort to all drift by, and I am just feeling about for the last time with the spike end of the spade, when I again hit off the hole and, opening it out, come upon a nice warm rat's nest made of leaves, with the ferret curled up snugly with a dead rat.

"Home, dogs home! Cheer up, Jack! Cold are you, and wet? Well, never mind; only two miles, and we will walk fast. Pepper, Pepper, Wasp, Wasp, where on earth have you got to? Ah, there you are, and a nice mess you have made of yourselves trying to scratch out a hole five hundred yards long. Come along all!" And off we tramp, Jack and I in the middle of the road, splish splash at every step, the water squirting high up our gaitered legs, and the dogs, with drooping tails, dripping coats and woe-begone looks, coming along behind us in Indian file close under the shelter, such as it is, of the hedge.

We pass the postman, who only nods, and meet a flock of sheep all draggled and dirty. An empty cart with a sack over the seat stands at the pot-house, and pigs wander listlessly about the yard with their backs arched up. Under the wagon-shed some cocks and hens stand each on one leg, with their tails drooping, apparently too disgusted to prune their feathers and fly up to roost in the rafters. The smoke beats down from the chimneys and gets lost in the wind and rain which buffets and pelts at our back. Cold spots begin to be felt at the bend of our arms and knees; then a shiver runs down the back, which develops into a trickle of water that at last gets into our boots and goes squish,

squish, at every step, and at last oozes over the tops; and our teeth chatter with cold, for now here and there among the rain-drops appear a few flakes of snow, which rest on the mud of the road for a second, and then melting, add to the deep slush that trickles down the hill by our side. At every open shed the dogs shelter a minute, shake themselves like dripping mops, and with arched backs stand on three legs and shiver; but we whistle them on and at last reach home. After throwing a good bundle of dry straw on the kennel benches and feeding dogs and ferrets, Jack and I get under shelter and soon find ourselves in dry clothes before a good fire, feeling a little swollen and stiff about our faces and hands, and much inclined for forty winks.

The wind howls in the chimney, lashes the bare branches of the trees, rattles the window frames, and appears angry that it cannot get at us, and the rain drives in fitful gusts against the windows, and hisses in the big wood fire on the hearth; and as I sit in my snug arm-chair, I dimly feel that the external storm adds greatly to the internal comfort, and then I fancy I nod off to sleep, for I think no more till supper is announced, and hunger and my wife stir me up to consciousness again.

Having finished a good supper and got my pipe drawing beautifully, I remember one or two things that I think the student should be told. The first is, never put a line on a ferret when *ratting*. It hampers a ferret in a narrow, twisting, turning rat's hole, and cutting into the soft earth at the turns soon brings the ferret to a dead stop. Then rats holes are chiefly in hedge-banks, which are full of roots, and the line is pretty sure to get twisted round some of these, and then it will be a long dig to free it. Remember, too, a ferret has to go down the hole and face a beast nearly as big as itself, with teeth like lancets and with courage to use them, and so should be as free as possible; and lining a ferret is about equal to setting a student with the gloves on to fight against another without them. Then some way back I mentioned ferrets' bells. They are little hollow brass balls with an iron shot in them that make a pretty tinkling sound, and are supposed to be tied round the ferret's neck. In my opinion, if you put a bell on it, you may as well put the ferret in the bag and keep it there. The theory about bells is, that a ferret running down a hole jingling its bell with fill a rat with fear and make it bolt, but this is all nonsense; rats are not so easily frightened. Again, it is said that if a ferret comes out of a hole in a thick hedge unseen, the bell will let you know where it is; but I must say I never lost a ferret in a hedge or felt the want of a belled one. I consider a bell a useless dead weight on a ferret, and the cord that goes round its neck to fasten it is apt to get hitched on to a root and hold the ferret a prisoner. A bell is only good for a sharp shopman to sell to a flat.[21]

I need hardly say, never muzzle a ferret when rat-catching. It would be brutal not to let the ferret have the use of its teeth to protect itself with. Muzzling ferrets appertains solely to rabbiting, but it is useful to know how to do it. Take a piece of twin a foot long, double it, and tie a loop at the double. Tie the string round the ferret's neck, with the loop on the top; bring the two ends down under the chin and tie them together there; pass them over the nose and tie them there, shutting the mouth tight; pass *one* along the nose, between the eyes, through the loop on the top of the neck, and bending it back, tie it to the other loose string from the knot on the top of the nose. Cut the ends off, and, provided you have not made a lot of "granny" knots, your muzzle will keep on all day. There are other ways of doing the trick, such as passing the string behind the ferret's dog-teeth,[22] bring it under the jaw, then over the nose, on the top of the neck; tie it there and again under the neck. I hate this plan, and have seen a ferret's mouth badly cut by the string. I have heard of another plan which is too brutal to mention. Cut the muzzle off

---

[21] That is to say "a fool".

[22] The canine or large front teeth.

directly you have done with it, for I don't suppose a ferret likes having its mouth tied up any more than you or I should.

Never wantonly hurt any animal, especially those that work for you and suffer in your service. Just think of the amount of pluck a ferret shows each time you put it into a rat's hole. Fancy yourself in its place, going down a lot of dark crooked passages that you don't know, only just wide enough to allow you to pass, and have to face a beast somewhat like yourself and as big, that you know will attack you. Why, if ferrets got VC's [Victoria Cross], they would, on high days and holidays when they wished to display them all, have to employ a string of sandwich-men[23] walking behind them with the boards covered with VC. Three or four times in my life I have had ferrets die of the wounds they have received from rats. I have had them in hospital for weeks, and I have had them blinded. Speaking of blind ferrets, I am not much of an oculist, but I don't believe a ferret can see in the dark. I never could find any difference between the way my blind ferret worked in a hole and that of one with good eyes; in fact, my blind ferret was a good a little beast as ever killed a rat, and she did kill many score after she lost both eyes. I believe a ferret when in a hole uses a sense we don't possess-I mean the sense of touch with the long nose whiskers.

Some years ago the *Field* opened its pages to a long discussion on the subject of ferrets sucking the blood of their victims after they have killed them.[24] Writers pretending to know all about it said they did do so. These men are to be pitied, not laughed at, for you see in the days of their youth "Rat-catching for the Use of Schools" was not written, and therefore they had not learnt better. A ferret no more sucks blood of the things it kills than a dog does. If you doubt this, give a fresh-killed rat to a ferret, let it fasten on it, and then peep at the corners of its mouth, and you

---

[23] Men with 'sandwich boards' front and back.

[24] The Field is very probably the oldest sporting magazine in the world, dating back to 1853. It is still being published today.

will find an opening there into the mouth, out of which blood would flow if the ferret had it in its mouth; and look down its throat, you will not find blood in it, nor will there be blood on the portion of the rat that has been held in its mouth. No, people are misled by a ferret sending its teeth deep home in the flesh and making a sucking sound as it with difficulty breathes through its nose and the corners of its mouth. If you watch a ferret after it has killed a rat, it will, as soon as it is sure the rat is dead, begin chewing at the skin of the head or throat till it has made an entrance, and will then eat the flesh.

To finish this chapter, I will tell you a story which you are never to put into practice. Some long time ago I found myself far from home in a country village, and having nothing to do, I went for a walk, and soon came upon a brother professional rat-catcher; and thinking I might learn a wrinkle from him that would come in useful, I joined him and carefully watched him and his dogs, I saw at once that three of the latter were very good and up to their work; but there was a fourth, a nondescript sort of beast with a long tail; that appeared quite useless; and I observed with amusement that directly the man put a ferret into a hole, the dog tucked its tail tight between its legs and went and stood well out in the field. I asked the man why he kept such a useless beast, and with a chuckle he answered, "Well, mate, I'll own up he ain't much to boast on for rat-killing, nor yet for looks, but he has his use like some other of we h-ugly ones. You see, sir, I've got one or two ferrets as won't come out of a 'ole, but stand a peeping at the h-entrance and waste a lot of time. Then that 'ere dawg comes in useful. I catches him, lifts him up, and sticks his bushy tail down to the ferret, who catches tight hold, and draws it out. Nothing ain't made for nothing, and I expect that dawg was made for drawing ferrets." The man may have been right, but I was quite sure the unfortunate dog did not take an active pleasure in his vocation.

There, young gentlemen, if you have well digested that chapter and forgotten the story at the end, you can put up your books and form up for you usual walk to the second milestone and back again; but before leaving, let me point out to you, Croker, minor, that if that caricature I have observed you drawing behind your book is meant for *me*, it is, like most things you do, incorrect; my nose is not so long, and I part my hair on the left side, not the right.

7

RAT-CATCHING and rabbit-catching are two distinct professions, but the greater part of the stock-in-trade that serves for one will answer for the other, and it is as well for the professional to be master of what I think I may call both branches of his business. A rat-catcher who did nothing but kill rats and refused a day's work with the rabbits would be like a medical man who would cut off limbs but would not give a pill, or a captain of a sailing-vessel who would not go to sea in a steamer; besides in these days it is the fashion to jumble up half a dozen businesses under one head and name. Just look at what the engineer does. Why, he is nowhere if he is not (besides being ready, as the engineer of the old school, to make railways, etc.)[25] a chemist, an electrician, and diplomat, a lawyer, a financier and a contractor, and even sometimes an honest man. If you are not in the fashion you are left behind as an old fogey, and so in this chapter we will discuss the art of rabbit-catching; and I trust all schoolmasters will furnish you, their students, with the opportunity of putting in practice in the field what you learn from this book at your desks.

Well, now for the requirements. We have got the dogs, we have got the ferrets, spade, bag, etc; but for rabbiting we must have a much more costly stock-in-trade if we are to do a big business. We shall require an ordinary gardener's spade for digging in soft sandy ground, where the rabbit burrows sometimes go in for yards, and as much as ten feet deep down; also another spade, longer in the blade than our ratting one, the sides more turned in, and with a handle ten feet long, with a steel hook at the end

---

[25] Indeed, this is the real trade of Henry Barkley!

instead of a spike.[26] With this spade we can sink down many feet after the hole is too deep for the ordinary spade, and the turned in sides will hold the soft earth and allow you to bring it to the surface. If you dig down on the top of a rabbit—as you will do when you know your work—the hook at the end will enable you to draw first it and then the ferret up by the string. We must have a piece of strong light supple cord, marked by a piece of red cloth drawn through the strands at ever yard, so that one can tell exactly how far in the ferret is; and it is as well to have a second shorter cord for work in stiff heavy ground, where the holes are never deep.[27] Next, we must have two or three dozen purse-nets, which are circular, about two feet in diameter, with a string rove round the outside mesh fastened to a peg. These are for covering over bolt holes to bag a rabbit when driven out by the ferrets. The nets should be made of the very best string, so as to be as light and fine as possible. The mesh should be just large enough to allow a rabbits head to pass through.

Like the postscript to a lady's letter, the chief item I have saved till the last, and I fear it will be some time before the ordinary rabbit-catcher will be able to afford it. I refer to long nets, which are used for running round or across a piece of covert to catch the rabbits as they are bustled about by the dogs.[28] A rabbit-catcher in full swing should have from eight hundred to a thousand yards of this, for with a good long net he will often kill as many rabbits in a few hours as he could do with the ferrets in a week.

I myself keep no special dog for rabbit-catching, chiefly because I have a neighbour who will always let me have a

---

[26] I have not heard of a shovel with such a long handle being used by rabbiters in Australia.

[27] Attaching lines to ferrets is rarely done in Australia, as is explained in the Introduction.

[28] See discussion on long nets in the Introduction.

cunning old lurcher[29] that he keeps, which is as good as gold, and as clever as a lawyer, and desperately fond of a day with me and my dogs.

I have three male ferrets, real monsters, strong enough to trot down a burrow and drag five or six yards of line after them with ease.

Having described all the tools, etc, necessary for work, I will now jot down, as an exercise for you students, a nice easy day's rabbiting that actually took place a few weeks ago—a sort of day that quite a young beginner might work with success. There had been a sharp rime[30] frost in the night, which still hung about in shady spots at eight o'clock in the morning, as Jack and I marched off with my dogs and ferrets, accompanied by Fly, the lurcher. By nine a.m. we began working field hedge-rows and banks, where rabbits were pretty plentiful and had been established for years in every description of burrow. There had been a lot of partridge and other shooting going on over this farm for the last month, and most of the rabbits had got a dislike to sitting out in the open, and were under ground, so we began at the burrows at once, the dogs driving every rabbit that was sitting out in the hedge back to their burrows as we walked along. We began work in a stiff clay bank far too hard for the rabbits to make deep holes in, and here we got on fast. I took the ditch side—in fact, I took the ditch itself—with a big ferret with a short line on, and I ran it into each hole I came to. Jack on the other side looked out for the bolt holes, and always laid down a little to one side, as much as possible out of sight, but with a hand just on the bank over the hole ready to catch a bolting rabbit. Fly and the other dogs took

---

[29] The Australian equivalent would be something like a 'Kangaroo Dog' – a very large dog of uncertain lineage but showing characteristics of both a Greyhound and an Irish wolfhound.

[30] Rime is a rough white ice deposit which forms on vertical surfaces exposed to the wind. It is formed by supercooled water droplets of fog freezing on contact with a surface it drifts past.

charge of the other holes, and all kept as quiet as possible.

In went the ferret, slowly dragging the line after him till I count two yards gone by the red marks on the line; then there is a halt for half a minute, then a loud rumbling[31] and the line is pulled fast through my fingers. Jack moves quickly, and the next instant a rabbit is thrown a little way out into the field with its neck broken.[32] Jack says, "Ferret out," then picks it up, draws the line through the hole, passes the ferret over to me, and we go on to the next, having filled up the entrance of the hole we have just worked. Hole after hole we ferreted much in the same way. Sometimes Jack bagged the bolting rabbit, sometimes the dogs, and now and then one bolted and got into the hedge before it could be caught and went back, but it was little use, for the dogs with Fly at their head were soon after it, and in a few minutes Fly was sure to have it, and would retrieve it back to Jack.

As we worked round a big field, we got into softer ground, a red sand and soil mixed; and here the holes were much deeper and often ran through the bank and out for yards under ground into the next field. Here Jack and I changed places, Jack doing the ferreting, and I going to his side with the garden spade. One, two, three, four, five yards the ferret went and stopped, and all was quiet. I listen, but not a sound. Jack pulls gently on the line and finds it tight, and for a minute we wait, hoping a rabbit may bolt from the hole the ferret went in at. But no such luck. I take the small ratting-spade, and with the spike end feel into the ground at the foot of the bank, and at once come upon the hole; this I open out and clear of earth, and Jack, who has crept through the hedge, kneels down and finds the line passing this hole in the direction of the field and going downwards. At that moment there is a sound like very distant thunder, and the line

---

[31] In Australia, this rumbling is more commonly called 'bolting' whereas the English 'bolt holes' are referred to as 'pops'.

[32] Dislocating the neck of the rabbit causes rapid death and is generally regarded as a humane way to kill a rabbit. The technique is explained later in this chapter.

is pulled quickly four yards further into the hole, and the marks show six yards are in. I go about this distance out into the field, lie down and place my ear close to the ground. I shift about in all directions listening intently, and at last hear a faint thudding sound. I shift again a few inches in this direction, and lose it; in that, and recover it; again a few inches, and the sound is directly under my head, but pretty deep down. I take the big spade and open out a hole a yard square, and dig down as far as I can reach. I get into the hole and sink deeper. I have to enlarge it a foot all round to get room, and then I dig down again till only my head appears above ground when I stand up. Then I take the long spade, and with that sink two more feet, and plump I come on the top of the hole, and the ferret shoves a sand-covered head up and looks at me. I reverse the long spade and catch the line with the hook and pull the ferret up, and then calling Jack, I send him head first into the well-like pit, holding on to one of his feet myself as I lie flat on the ground to allow him to go deep enough. In a minute a dead rabbit is taken out and two live ones, whose necks Jack breaks as he hangs suspended, and then I pull him up with his plunder, and he rights himself on the surface, very red in the face, very sandy, spluttering and rubbing his eyes. Then the ferret is swung down again by the line, it goes a little way into the hole and returns, and so we know we have made a clean sweep. The big hole is filled up and stamped down, and after filling a pipe and resting a few minutes, on we go with our work.

On the high sandy part of the field we have several deep digs like the above, with varying success, and we rejoice when we reach the last side of the field and get into clay again, where holes are short and most of the rabbits bolt at once. During all the day we stopped once for half-an-hour to get a snack of bread and cheese, and by the time the cock partridges began to call their families together for roost, and the teams in the next field to knock off ploughing, we are all, man, boy, dogs and ferrets, fairly tired, and are glad to tumble seventeen couple[33] of rabbits into the keeper's cart that has been sent out for them, and trudge off home ourselves.

Now for another day's sport that was quite different. No dogs with us, only a bag of ready-muzzled ferrets, a bundle of purse nets and a spade. Success will depend on perfect quiet, and even the patter of the dogs' feet would spoil our sport, so they are at home for once, and Jack and I are alone. It is one of those soft mild dull days that now then appear in mid-winter, a sort of day to gladden the heart of foxhunters and doctors, and to make wiseacres shake their heads and say "most unseasonable." It is a good day for Jack and me, and we feel confident as we steal into a plantation of tall spruce firs, placed so thick on the ground that beneath them is perpetual twilight, and not a blade of grass or bramble to hide the thick carpet of needle points. Softly we creep forward to a lot of burrows we know of in the corner of the wood, and then I go forward alone and spread a net loosely over every hole, firmly pegging it down by the cord. This done I stand quietly down-wind of the holes, and Jack comes and slips the six ferrets all into different holes, and then crouches down on his knees. All is quiet; only the whisperings of the tree-tops, the occasional chirp of a bird, or the rustle of a mouse in the dead leaves. Five minutes pass, and then out dashes a rabbit into a net, which draws up round it. Jack moves forward on tip-toe, kills the rabbit and takes it out of the net, and covers the hole again. While he is doing this, three more rabbits have bolted and got netted, one has escaped, and a ferret has come out. The captured ones are killed, the ferret sent into another hole, and for an hour this work goes on, and during all the time neither of us have spoken, for we know there is nothing that scares wild animals more than the human voice, unless it is the jingle of

---

[33] In Australia the term 'couple' is replaced by 'pair'. Dead rabbits are 'paired' by interlacing the back legs in a particular fashion, making a small slit between tendon and bone.

metals, such as a bunch of keys rattling. They dread the human voice because they have had too much experience of it, and the rattle of metal because they have not had experience enough of it, for it is a sound they have never heard, and nothing like, in the quiet woods and fields. On the other hand, animals pay but little attention to a whistle, for in one shape or another they are constantly hearing it from feathered companions.[34]

But to go back to our netting. An hour over, we pick up the ferrets as they come out and bag them, and then I go off to some fresh holes and spread the nets again, and we repeat the same performance; and during the day we kill, without any digging or hard work, about twenty-two couple of rabbits. In the above account I have written of a day's sport that took place in a fir plantation in a little village in Norfolk,[35] where it would have been madness to work the ferrets without muzzling them, for they would have been sure to kill some rabbits in the holes and then have laid up; but I should mention that I have killed many rabbits in the same way on the Cotswold Hills in Gloucestershire, and I was much astonished when I first got there to find men who thoroughly understood their business working their ferrets under nets without muzzling them. I adopted the plan myself, and have rarely had a ferret kill a rabbit underground. For some reason that I could never find out, a Cotswold rabbit will always bolt from a hole with a ferret in if it can. It is well known in Norfolk that if a rabbit is run into a hole by dogs, you may ferret it if you like, but it will never bolt, and it must be dug out. But in Gloucestershire I have seen the same rabbit bolt out of a hole, get shot at, be run by dogs, go to ground, and again bolt at once from a ferret. Few professionals ever use a line on a ferret on

---

[34] Given the use of a whistle in directing sheep dogs, this is a somewhat doubtful assumption.

[35] The discussion here and below strongly suggests that the author, if not a professional ratter/rabbiter at some stage in his life, certainly writes from personal experience and not second-hand.

the Cotswold, one reason being that the burrows are nearly all in rocky ground, and there would be danger of the line being caught in the numerous cracks; besides it is not required, for a rabbit there is sure to bolt, and for this reason it is twice as easy to kill rabbits in Gloucestershire as it is in Norfolk, especially in the sandy or soft soil of the latter county.

Let me here beg of all my readers, especially students, never to keep a poor rabbit alive in their hands a second. I don't suppose any who read this book could be so unsportsmanlike and brutal as to keep a rabbit alive to course and torture over again with dogs, or for the fun of shooting at the poor little beast. Such ruffians should never be allowed a day's sport on a *gentlemen's* property. They are only fit to go out mole-catching. No, directly you have a live rabbit in your hand, take it by its hind legs with your right hand, and the head with your left, with two fingers under its face; with these fingers turn the head back, and give the rabbit a smart quick stretch, and in an instant all its sufferings are over. Never hit it with your hand or stick behind the ears; first, because you are not quite sure to kill it with the first blow; and secondly, if you do, half the blood in the rabbit will settle in a great bruise at the spot where it was struck, and make that portion unfit for table.

That is sufficient for this morning, and you may now turn to a little lighter work with some algebra.

8

FORTUNATELY I don't live by the sea. I say fortunately, because I look upon the sea as a swindler, for it robs one of just half one's little world and upsets all calculations by forcing one to live in a mean semicircle. I actually know a rat-catcher who is stupid enough to live in a village on the east coast, and half his time he and his dogs are at home in idleness and are half starved, because the ever-restless tiresome sea rolls about and disports itself over all that is east of the village, so the poor man can only go rat-catching in one direction. Now and then I go to the seaside, but when I go there it is on business—not in my Sunday clothes and with a "tripper's" return ticket, but with my dogs, ferrets, nets (the long ones) and the boy Jack; he and I dressed in our well-worn corduroys, gaiters, and navvy boots; and instead of choosing a town to visit with Marine Parade, Esplanades, Lodgings to let, Brass Bands, Minstrels and spouting M.P.'s, we go to a little village unknown to "trippers," and put up at a small inn for a week or ten days. We sleep in a room not unlike a hayloft, and take our meals and rest in the common kitchen, with its rattling latticed windows and sanded floor.

We go there twice each winter to kill rabbits on what are called the "Denes,"[36] which are great, wide, down-like lands on the top of the steep earth cliff, partially covered with the everflowering gorse, a cover dear to rabbits and all sorts of game. We reach the inn in time for an early dinner; and after we have housed the ferrets in a big tub and the dogs in a warm dry shed

---

[36] There is an area in today's English called "The Denes". It is part of Darlington in the north-east, but bears no similarity to the country described by Barkley. For one thing, it is not coastal. In Suffolk, there is a "Denes Beach" and perhaps this might have some connection to Barkley's spot.

with heaps of straw to sleep on, Jack and I despatch our food
and then start off to inspect the field of our future operations.
We have not far to go. First down the street, past two or three
dozen flint-pebble cottages; past the church, with its square
tower so high that it makes the really big church look small in
proportion; past the rectory; past the schools, where some forty
or fifty future fisherman and sailors have just finished their tasks
for the day and come rolling out, dressed all alike in dark, sea-
stained, canvas trousers and thick sailor jerseys; past the low
one-storied cottage where the old retired naval captain has
lived for many years, and then up a sandy lane between high
crumbling banks and out on to the open Denes. We take a path
that runs close along on the top of the cliff, mounting a steep hill
as we go till we reach a spot half a mile further on, where the
sea cliff is four hundred feet high and nearly perpendicular; and
here among the ruins of an old church, part of which has fallen
with the slipping cliff into the sea many years ago, Jack and I
halt and take a look round. We are on the highest spot within
miles, and spread out in front of us, as we face inland, are, first,
the down-like hills, dotted over with patches of gorse and with
turf between as fine and soft as a Persian carpet;[37] then cultivated
fields intersected by thick hedges; and in the distance we could
distinguish a clustering village here, a homestead there, an old
manor-house in its well-kept garden and park-like grounds, and
in all directions the square, solid, picturesque towers of village
churches peeping from among the trees, that became thicker and
thicker the further the eye travelled from the sea. Close to our
left, just under the shoulder of a hill which protects it from the
keen east wind off the sea, is a tiny village of some ten cottages,
all different, all neat and snug-looking, each in its own garden.
There is a stand of beehives in one, a honeysuckle-covered
porch to another, and, though it is mid-winter, there is a warm

---

[37] Rabbits in large numbers often selectively graze certain patches and maintain them as
a fine lawn, preferring the soft, new shoots to higher, rank vegetation.

home-like look about all. Then there is the one farm-house, well kept and well cared for, but old and belonging to other days, as its gables and low windows denote; and from our high hill we look over the house into a garden and orchard beyond, both enclosed by grey lichen-covered walls. On either side in front of the house are the farm buildings, all, from the big barn to the row of pigsties, thatched with long reeds which give the whole a pleasant English home appearance.

There are big yards filled with red and white cattle up to their middle in straw, others full of horses or young calves; cocks and hens are everywhere, ducks and geese swim in the big pond by the side of the road, and turkeys, so big and plump they make one long for Christmas, mob together in the yard, and the turkey-cocks "gobble-gobble" at a boy who is infuriating them by whistling. A man crosses the yard with two pails on a yoke, evidently going a-milking; and another passes with a perfect hay-stack on his back, and a dozen great heavy horses come out of the stable in Indian file and stump off to the pond to drink. Beyond the farmstead, in a field on the right of the road, is a double row of heaped up mangels and swedes;[38] and a little further on are a number of stacks, so neatly built and thatched that it seems quite a pity they should soon be pulled down and thrashed, but all showing signs of prosperity and plenty.

Beyond this stands a tiny church, with reed-thatch roof. It is all, church and tower, built of round flint stones as big as oranges, cleverly split in two and the flat side facing outwards; and from the dog tooth Saxon arch over the door one knows it has seen many generations pass away and find rest from the buffets and storms of the world in the peaceful, carefully-tended "God's acre" that surrounds it.[39] If one passed down the red gravel churchyard path, and on in front of the south door to the

---

[38] Root crops used for animal feed. The mangel is a type of beet, the swede a type of turnip.

[39] Cemeteries were once commonly referred to as "God's Acre".

far corner, under the big cedar, a small door would be found, which would lead through a well-kept, old-fashioned garden to the Rectory; a good old Elizabethan house, covered with thick creepers up to the very eaves, the model of one of England's snug homes—homes that have turned out the very best men the dear old land has produces, to fight, struggle, conquer or die in all professions, in all parts of the world; men who in such shelters learned to be honest and true, brave and persevering, lions in courage, women in gentleness; who could face hardships and poverty without a moan, and prosperity and riches without swagger; and through all the difficulties of life thought of the old home, and when success arrived, be they ever so far way, packed up and came back to finish their days in just such another home and such surroundings.[40]

Turn round now, Jack; turn round and take a look at the restless sea rolling its big waters on the smooth strip of sand there below *on this side;* and on the other, Jack, far, far way over there in the south, on the other side of the world, laving the roots of the palm and the mangrove, beneath the burning rays of tropical suns; and away round here, Jack, far in the north, dashing its storm-driven waves against the face of frost-bound rocks and treacherous icebergs. There on the dancing waters, with all sails set, chasing the lights and shadows as they flit before it, sails a boat bound south to sunny climes. There on the horizon, against wind and wave, steams a collier, taking fuel to lands where the snow lies deep on the ground for four months in the year; and right and left, outward bound or coming home, are various white sails dotting the waves. But, Jack, how about supper? I ordered eggs and bacon for supper, and those chimney corners at the inn looked as if they might be snug and warm to smoke a pipe in afterwards before turning in. Step on, Jack and have supper

---

[40] We may wonder whether Barkley himself may have retired in such a place. His careful descriptions and emotional attachment to the scene he describes, is very evident here. The following paragraph is also suggestive of a well-travelled man.

ready in half an hour, while I go round by the Rectory and see if the two young gentlemen are at home. They are the right sort, and as keen as Pepper after the rabbits, and they always have half a dozen good terriers as fond of the sport as they are.

At the Rectory I received a kindly welcome from Miss Madge Ashfield, the rector's only daughter and the sister of the two lads I came to enquire for; and I was told that they were not yet back from school, but were expected three days, and that only that morning a letter came from them asking when I was likely to come and work the Denes. I comforted Miss Madge, who at first feared the pick of the sport might be over before her brothers arrived, by telling her that for the next four days Jack and I should be busy "doctoring" holes, and that during this time we could not "away with" boys and girls or dogs, as both were too noisy for the work.

Miss Madge took me round the kennels to see some rough wire-haired terriers, old friends; also three new ones, all supposed to be wonders; and she told me she would arrange for her brothers to bring one day five small beagles belonging to a friend.

Jack and I did our duty by the ham and eggs that night at the inn, and the pipe in the old-fashioned chimney corner was very sweet; and if the beds were a bit hard and knubbly, we did not keep awake to think of them, for we had both been up since daybreak.

By eight o'clock the next morning we had finished breakfast, given the dogs a few minutes' run to stretch their legs, fed the ferrets that were not wanted, and were on our way to the Denes, each with two strong male ferrets, a spade, and game-bag with cold meat and bread in it. We were on our way to "doctor" the burrows, and this is done by running a muzzled ferret that has first been smeared with a little spirits of tar[41] down every hole,

---

[41] See footnote 8 on page 28

with a line on it. It is necessary to keep very quiet, so as to get the rabbits to bolt. We don't want to kill a single rabbit, but only to disturb hole after hole, bolt what rabbits we can, and leave a nice sweet smell of tarred ferret behind us. No time lost. Jack goes one way and I another, and every hole is visited till evening shades stop us; then back home to supper and bed, and at it again in the morning; but on the second day we begin by visiting each hole we ferreted the day before, stopping them tight down with sods, and sticking a piece of white paper on the top of such stopped holes. No fear of shutting in a rabbit, as the smell of the tarred ferret will keep them out for days; and no fear of their opening the stopping, as the paper will drive them away. For four days this work goes on, and we are ready to wager there is not a hole in the cliffs or Denes that is not doctored, and not a rabbit that is not above the ground.

It was Wednesday night when we had finished, and that evening the two boys from the Rectory came down to the inn to see us and get instructions for the morrow; but I was glad they did not stay long, for we wanted to go to bed early, so as to get a good night and yet be up betimes. By eight o'clock next morning, Jack and I were already back from the Denes, after having run out one thousand yards of long nets. The nets are in lengths of about one hundred yards, and two feet six inches high, made of fine string, an each of the top and bottom meshes knotted on to a cord that runs the entire length. To set these nets, they are threaded on to a smooth stick, four feet long, and the stick with the nets on is thrown over a man's shoulder. The man walks off with the nets along the border of the piece of ground to be enclosed, while another, after fixing the end of the first net fast to a starting stick, follows behind. As the man with the net proceeds, he lets the net slip slowly off the stick on his shoulder, piece by piece; and, as it come down, the man behind picks up the top line, gives the net a shake, and twists the line round the top of stakes previously placed in the ground about fifty yards apart, taking care as he goes that the bottom of the net lies for a few inches on the ground. In this way squares of gorse of about two hundred yards can be entirely enclosed, and every rabbit inside them surrounded like sheep inside a fold.

Our breakfast over, we were soon out again with all our dogs (except old Chance, who had been left at home on account of her age, and also on account of her trick of always liking to go up to the carrier's each night to sleep), and we had also two real good lurchers. At the foot of the Denes we met the boys from the Rectory, with a friend about their own age, and the curate of the next parish with a business-like ash stick under his arm; and among them they had mustered a pack of ten terriers, some of which wanted to begin work by a fight with my dogs; but it takes two to make a quarrel. And my dogs knew better than to waste their strength in fighting when there was a day's work in front of them.

In a few minutes we were at the first piece of netted gorse—a real tearer, close, compact and a mass of thorns; but what dogs or boys care for gorse thorns when rabbits are on foot? So it is, "over you go, boys!" "Hie in dogs! Roust them out there!" and the old dogs spring the nets and are at work in a minute, while the young ones blunder and struggle in the nets, and have to be lifted over. The curate, Jack and I, and the man who drove the cart with the nets, and who will carry off the dead rabbits, stand at the nets and take out and kill the rabbits that get caught; and for the first hour we have as much as we can do, and work our hardest. Many rabbits do get through the nets, and others go back, and these latter it is difficult to get into the nets a second time, and they are killed by the dogs in the thick gorse. Yap! Yap! Yap! "Hie in, good dogs! Hie in, young ones! Ah! back there! back! No going over the nets! Would you? Look here! Hie there! In you go!" Yap! Yap! Yap! All scurry, rush and bustle; and the Rectory boys and their friend are all over the square at once, and in ten minutes so tingle from innumerable pricks from the gorse that they are benumbed and feel them no more. "Go, Fly, go!" and a big hare dashes out, with Fly after it, and both jump the net and make for another clump of gorse, but Fly has never been beaten since she was a puppy, and soon returns with the hare in her mouth." "Hie in, dogs! Hie in!" There are more yet, and we are bound to make a clean sweep; and so the work goes on.

First one patch, and then another, till lunch-time, which said lunch, according to a long-standing custom, comes up in a cart from the Rectory; but after snatching a hurried bit, the man and I have to bustle away to shift the nets, a work that keeps us hard at it for an hour and more; but long before we have done, the boys, parson and dogs are at it again in one of the first patches we have surrounded, and it is night and the moon is up before we have finished and picked up the nets. We find on counting the bag that we have two hundred and seventy rabbits, and feel content with our day's work. On Friday and Saturday the same work, and when we turned homewards on this last night, it was as much as man, boys or dogs could do to drag themselves along; but we had killed six hundred and fifty rabbits in the three days and were well content.

9

SUNDAY was to us all a real day of rest, and we enjoyed every minute of it, and for once listened to a very long sermon without the fidgets. The Rectory boys came up for a chat in the afternoon, so we let the dogs out and went down to the beach and strolled quietly about, neither dogs nor humans indulging in anything like play—all were too stiff and sore to think of it.

We were all out again early on Monday morning, but without nets and taking only sticks; and we spent a short day, with a long lunch, looking up outlying rabbits in the hedges of the farm at the foot of the Denes; and here the two lurchers, who during the days at the nets had taken it easy and refused to face the gorse, had the chief of the work, for directly a rabbit was started by the other dogs, it made straight off across the open for the gorse on the Denes, and the lurchers were the only dogs fast enough to catch them. We finally had to give up work because the dogs of all sorts were too tired to move, and also because the weather, that had been fine and calm all the previous week, began to break, and before we reached shelter there was half a gale sending big green waves thundering on to the beach and carrying the salt spray far inland.

That night, after Jack was in bed and asleep, I put on my hat and went out, called by the noise of the waters. I joined a group of weather-beaten hard-featured men dressed in thick blue jerseys and "sou-wester" hats, who stood with their hands tucked deep into their trouser pockets, watching the sea from behind the shelter of a boat stranded high up on the beach. I got a civil word of greeting as I came up, and then we all watched in silence, for by this time the "half gale" had become a storm, and it was only by shouting we could have made each other hear. It

was a wild weird scene, awe-inspiring, but intensely attractive—
at least *I* found it so; but then such scenes did not often come
before me, and I daresay my companions, who were well used
to being out on such a night, only felt thankful they were safe
on shore, and thought with anxiety of those of their friends and
neighbours who were out battling with the storm. The moon
when I reached the beach was nearly at the full and high up in
the heavens, but it shed a fitful light, as each few seconds dark
clouds and veils of mist flew across its face. One moment the sea
lay before us a dark black mass, only marked along the beach
by a broad strip of breaking, foam-crested waves; and the next
it was dancing, tossing, roaring sheet of ever-changing liquid
silver; or far way we would see the spray like pearls rising high
in the air before the storm, and at our feet the waves curled up
like huge furious monsters, dashing at the sands and shingle as
if bent on destruction, and then with a swirl sliding back, a mass
of foam, to meet and join the next wave, and with its help again
come on to the attack.

Over and over again I fancied I could hear the shrieks and
groans of people in distress, and I turned for confirmation of
my fancies to the faces of my companions; but all remained
unmoved, but bore the quiet determined look that assured me
that, had any unfortunate beings called for help from the midst
of those wild waters, at the risk of those men's lives it would
unhesitatingly have been given. Once for a moment, when a thin
mist swept before the moon and made the light on the waters
appear more like day than night, I clearly saw on the horizon the
upper part of a ship's masts, with some sails bent to their yards,
and all heeled over as if the ship were then about to founder,
and I gave a loud exclamation; but an old sailor put his hand on
my shoulder and called in my ear, "All right, master, all right!
We have watched her for a quarter of an hour trying to make the
point of the sands yonder, and she is now past them and has an
open sea. She is as safe as you are now, thank God; but it was a

near shave and we thought she and all in her were gone." Often since then in my dreams I have seen that wind-tossed sea, and heard the roar of the waters and the screams of the storm, and seen those masts and sails heeling over, and have awoke with a start and dread fear in my heart.

I had been tired when I came in from work, and I had a snug warm bed waiting for me, and moreover I reasoned that watching a storm in the dead of night was no part of a rat-catcher's duty; but I was so fascinated I could not tear myself away, and I stood with my companions behind the boat till long after midnight. Then two other figures dressed like my companions joined us, and it was only when they spoke that I recognised one as the parson of the parish, and the other as the young curate who had helped us with the rabbits. Both asked a few questions of the sailors, who seemed eager to give them information; and then the rector, turning to me, said; "You will be perished by the cold if you stand here longer. Come with me, and I will show you a picture of a different sort, but yet one that I think will interest you." I readily accepted and followed my friend, who, though far from a young man, bore the buffeting of the storm manfully; and he led me up through the village street, and then turning down a short steep lane brought me a little cove that was partly sheltered by a spit of rock that jutted out into the sea. There, such as it was, was the harbour of the village, and by the fitful light I could see some dozen fishing boats drawn up high on the beach above the force of the waves; and beyond, cluster of low, one-storied cottages and sheds, with small boats, spars, timbers, windlasses, etc., all denoting the home of fishermen. From this cove, early that morning, two boats had sailed with their nets for the fishing grounds out beyond the sands, and it was for these my friends behind the boat were patiently watching, and it was to say a few words to cheer and comfort the wives and families of these men that the old rector had now come.

From a latticed window just in front of us a bright lamp shed its rays over the cove, and the rector took me straight to the door of this house, and having knocked and been told to come in, he lifted the latch and ushered me inside. The room was like hundreds of others along the coast, the homes of the toilers of the deep, and bore evident signs of being made by men more used to ships than stone or brick buildings. It was a good large room, very low, with heavy rafters overhead, which, with the planks of which the walls were constructed, had doubtless been taken from boats and ships that had served their time on the sea. The open fireplace at the end, with its wide chimney, was the only part of the building not made of old ship timbers and planks, and there was a strong smell of tar from these and from sundry coils of dark rope that were stowed away in a far corner. The long table down the middle of the room was of mahogany and had seen better days in a captain's cabin. The benches round the walls had served as seats on some big ship's deck; and there were swinging lamps and racks hung overhead from the rafters, with rudders, boat-hook, snatch-block, belaying pins, and various things I did not know the use of; but all were neatly arranged. There was a large arm-chair made out of a barrel set ready by the side of the hearth, on which was spread clean flannel clothes to warm and air, in readiness for the home-coming of the wet and tired husband.

In front of the fire, attending to it and to three or four pots and kettles that simmered on the hearth, stood a woman about thirty years of age—just an ordinary fisherman's wife, strong and well shaped, without beauty of feature, but bright and intelligent looking; and when a smile lit up her face, it shed such a kindly ray that one felt that the husband in the little fishing boat on the storm-tossed deep might have his eyes fixed on the lantern burning in the window, but it would be the light of the wife's smile that kept his hand steady on the helm and guided the boat, and made him long to round the point and come to anchor.

On the other side of the hearth was another arm-chair, also

made out of a barrel, but much smaller; and in this, packed tightly and snugly round with cushions, half-sat, half-reclined a boy about ten years of age; but, alas! A pair of crutches leaning in the corner beside him at once told a sad tale. I know the points and beauties of all sorts of dogs, and always admire them, but I am not much of a hand at the good points and beauties of men and women, and as far boys, it is rare I see anything but mischief written in their faces; but somehow I could not take my eyes off the boy in the chair. I suppose because it was so different to any other young face I had ever seen, and so different to what one might expect to find amid the surrounding of a fisherman's cottage.

It was a dark, delicate, oval face, like a girl's, with finely cut features, and a complexion as fair as the petals of an apple blossom; but it was his great brown eyes and long eyelashes, black as night, that held the attention, together with a look of deep patient suffering, mingled with gentleness and love that lit all up, and filled even the heart of a rough old rat-catcher like me with a feeling of deep pity and an intense desire a protect a befriend a small creature who looked too fragile, too beautiful, and too good for this old work-a-day world of ours, and as if he were only tarrying for a short while before going to his eternal home, where his features will be beautified by perfect love, and will lose the look of suffering and pain.

The rector, taking off his "sou'-wester" as he entered, turned to the woman with a cheery voice, and said, "Well, Mary, how are you and the boy?—how are you, my man? I happened to be passing" (just as if it were quite a common thing for a parson to be out on the loose at one a.m on a winter's night), "and I thought I would just call in to say that the men at the boats tell me that the bark of this gale is far worse than its bite, and that it is a fair, honest, rattling gale that such good sailors as your husband care nothing for, and that we may expect the boats in with the daylight, so you may keep the pots boiling. But why isn't that youngster snug in bed and asleep? Oh ! He can't sleep when the wind howls, and Jack is away! Why, my boy, Jack will laugh at you when he comes home, and say he don't want such big, tired-looking eyes watching for him! Well, it will be morning soon, and, please God, Jack will be here, and will have popped you into bed himself before most of the world are up and about." At this Mary smiled; and the little boy, with a low laugh, said; "Jack knows Mary and I are waiting for him. Jack says he can often see us, and all we are doing, when he is out at sea in a raging storm, and the night is ever so dark; and he'd feel bad, Jack would, if I was not up to see him eat his supper; and besides, Mary could not sit here alone and listen to the wind and sea, and I am never tired and sleepy when waiting for Jack. Besides, Jack says he must tell someone all he has done and seen while he gets his supper, and Mary is too busy after the nets and things, so I sit here, and Jack tells me of such wonderful things; it is just lovely to hear him."

The rector would not sit down, and soon hurried me off to another cottage, much such another as the first; but instead of Mary and the boy, we found a great, tall, gaunt old woman, sitting up before the fire, waiting for her two grandsons, who were away in the same boat with Jack; but to the rector's cheery, hopeful words, the woman answered with a bitter, sharp, complaining tongue; "I don't want no stop-at-home idle chaps to tell me what a storm is. Danger! Who says there's danger? Danger with a little puff of wind like this? Not but what both of those boys will be washed ashore one day as their grandfather and father were. It's in the blood, and trying for a lone woman. Drat the boys! I told them not to go off with Jack. I could see plain for days that it was coming on to blow; but oh, no! They know better than me, who have lived to lose their father in such a storm as this, and to see his boat with my own eyes go to pieces on the Point as she came in, and not a man saved, and me left with them boys to keep. God only knows how I did it, and now they are that masterful they won't pay no attention to me." And then, as a hurricane of wind dashed at the door and windows and sent the smoke from the wood fire far out into the room, the poor old thing started and turned to the night outside with a look of terror; and, as the storm rushed on, and then there was a lull, she threw her apron over her head and sobbed for fear and deep anxiety for her grandsons.

The rector comforted her with gentle words and praise of her pluck and nerves; and as he and I returned to the beach, he told me that the old woman had once been the prettiest girl for many miles round, that when her boys were far too young to help her father had been drowned by the upsetting of his boat on the Point, and from that day she had worked and toiled, mending nets and selling fish in fair weather and foul, often weary and half-starved, but succeeding in the end to keep her old cottage over her head, and to bring her boys up respectably and turn them out two of the smartest fishermen along the coast.

As we left the cottage the first tender light of the morning was paling the eastern sky far out of the sea, and hastening on to the Point, we could just make out a distant sail appearing now and then out of the departing darkness of the night, and before half an hour was over the rector declared it to be Jack's boat coming in fast before the wind. All the village was a stir in a minute, old men and young women and children hurrying to the cove and making ready for the home-coming; and in a few minutes the boat, with Jack holding the helm and the old woman's boys sitting crouched low down, dashed past the Point, turned sharp into the cove, and down in a moment fell the sail and the anchor-chain rattled out of the bows. There was no cheering or noisy welcome or rejoicing, for such scenes were the daily incidents in the life of the village; but everyone lent a helping hand, and in a few minutes Jack and his men were on shore. The old grandmother was there, but took no notice of her grandsons, who marched off to the cottage laden with oars, etc., where the old woman had just preceded them to put out the breakfast.

The rector and I turned to go home, and as I passed the cottage where Jack lived I glanced in and saw him standing on the hearth, tall, massive, weather-beaten and rugged, with the lame boy high up in his arms looking hard in his face, and both man and child had such a happy contented smile on their faces that it did me good to see, and I think may have rejoiced even the angels above.

When parting from me at the inn door, the rector said that if I liked to step up to the rectory that evening after my supper he would find me a pipe of tobacco, and tell me all that was known of the history of the little boy who had awakened such an interest in me, for, he added, "it is a very curious story."

10

At eight o'clock, having fed my dogs and ferrets and left my Jack chatting in the harness-room with the rector's old coachman, I found myself in a snug arm-chair, pipe in mouth, my feet on the fender, and the rector sitting opposite me and his study, he also enjoying an after-dinner pipe; and after a chat over the events of the day and of the storm of the previous night, the rector began the history of the poor lame boy at the cottage thus:

I dare say you remember that about eight years ago the Irish question was giving the authorities much trouble and anxiety owing to the active turn it had then taken. Hideous murders were of daily occurrence in that unfortunate country. Dynamite was being used in London to destroy our public buildings, and many of our statesmen were being tracked by paid assassins.[42] Strict orders had been issued by the authorities to watch all our ports to prevent the landing from America of arms and infernal machines, and both the police and Customs officers were on the alert; and yet, in spite of all, bloodthirsty, cowardly dynamiters and assassins succeeded in sneaking into the country, and every now and then perpetrated some hateful outrage. Well, it was during this time that one November morning a queer-looking yacht-like vessel appeared in the offing, and for two day kept standing about. During the day-time it was well out in the offing, but once

---

[42] This bombing campaign was carried out by the Irish Republican Brotherhood, also known as the Fenians, from 1881 to 1885. I presume that "the Irish question" referred to here was the Fenian uprising of 1865 (and its aftermath), when thousands of men took part in a rebellion against British rule in Ireland. For their part, the Irish and Irish-American participants in this campaign regarded it as a legitimate use of force to gain liberty and independence. The British, of course, took a different view. The earlier 'Great Famine" in Ireland, between 1845 and 1852, had also turned many Irish against British rule, including Irish emigrants now living overseas.

or twice at night it was noticed by the coastguard and sailors to have come close in to land, and altogether its movements were so mysterious that our suspicions were finally aroused, and the officer of the coastguard telegraphed to the captain of the gunboat stationed at Brockmouth[43] to put him on the alert.

For some days after this nothing was seen of the yacht, and our suspicions were lulled, and life in our quiet little village had settled down to its usual routine, when early one stormy morning the strange vessel was again seen close off the land, and a boat manned by six men put off for the little harbour; and just as it rounded the Point and got into smooth water, a dog-cart, that we all recognised as one let out for hire in a town ten miles inland, drove down to the beach. Beside the driver sat a tall, thin, dark man, but the few people on the beach had only time to observe this and that he had the dress and appearance of a gentlemen, when he sprang from the cart and hurried to where the boat lay, and without hesitating a moment or speaking to anyone he waded out through the low surf to the boat, which at once left the harbour and made the best of its way to the yacht, which as soon as all were on board hoisted all sail and was soon out of sight, driven along by a storm that became in the course of the day as fierce a one as that of last night. There was much talk on the beach among the fishermen and in the village among us all as to what the yacht could be and who the stranger was; and we gathered from the driver of the dog-cart, who had put up his horse at the inn to rest, that he had been called by the porter at the railway station to drive the gentleman over; but that he had not heard his name, or what business brought him here. The driver, who was a sharp old fellow, said the gentleman had chatted with him as he came along, but kept pressing him to drive faster and faster, and gave him five shillings above his fare to use his best speed, and he added; 'I don't know who he is, or what his business may be, but I know one thing—he is an

Irishman. I can tell it by his tongue, and by his queer-looking blue eyes and dark hair.'

Four and twenty hours passed, and during that time many people, I among the number, did not go to bed, for the storm which had sprung up with the departing yacht had blown itself into half a hurricane, and there were fishing boats out, which made us all anxious. As we did last night, or rather this morning, I went round to a few of the fishermen's houses where there were anxious wives and mothers waiting for the absent, and chatted with and cheered them, and I was leaving the two cottages that I daresay you noticed close under the rock towards the Point when the first streaks of morning began to appear in the east. I love to see the day break at any time, but I especially like to watch it over a stormy angry sea; and therefore sheltering myself a little behind a boulder, I stood gazing for a while, when presently, like a thing of life, came plunging and driving from the very gates of the morning the same yacht that had so puzzled us. On and on it came, close-hauled to the wind, straight for the narrow rock-bound jaws of the cove; and I saw at a glance that, if it kept its course, it must strike on a group of rocks some half-mile out at sea; and, parson as I am, I knew, should she strike them, no human aid could save the lives of those on board.

I hardly know what I did, except that I took off my coat and waved it frantically, and mounted the highest pinnacle on the rocky point to make myself seen by the fated crew; but though at last I could actually distinguish two men at the wheel holding the vessel close to the wind, yet they took no notice, and came on and on, leaping waves mountains high one minute, and lost to sight the next in the trough of the seas. Scores of fishermen soon joined me, and even their wives followed and crouched near, behind the rocks; and so fully was the ship's danger realized, that from time to time a deep groan, half of despair, half prayer, went up from all. There was but one hope—could the yacht be kept close enough to the wind to lead those steering her to believe they could make the entrance of the harbour? Or would she be carried far enough to windward to make this impossible, and so force those in charge to alter her course to avoid the stiff cliffs beyond? Ah, no! We saw as we watched that she was too good a vessel to fall off the leeward, and those handling her too good sailors to allow her to do so, for she flew over the waves like a beautiful bird for the entrance of the harbour, and the sunken rocks were in her direct line!

Suddenly as we watched, with every sense strained to the utmost, and our eyes rivetted on the doomed ship, we heard away out to sea the boom of a big gun, and then another, and presently we saw emerging from the fast diminishing darkness a low, long steamer. At first we thought it was a ship also in deep distress, making signals; but the old sailors soon saw this was not so, and declared it was a gunboat firing at the yacht in the hope of driving her on to the rock-bound coast, and also to attract the attention of the coastguard, so that, should she reach the harbour, those on board might be prevented from escaping the hands of justice. It was a cruel service for British sailors to be employed on, however necessary, and hard to witness. Man hunting man to his death, when the wind and waves already held open the portals of eternity before him, and little short of a miracle could avert his doom!

A few minutes, a few hundred yards, and the yacht is on the rocks! Gallantly she glides along the side of that green wave and dashes the foam from her crest ere she plunges deep into the sea. A monster wave rolls fast upon her as if to swallow her quivering form. High, high she rises, till half her length is in the air over the crest of the wave, and then down she sinks; then the crash comes. Waves dash over her, her masts fall, her boats are wrenched from her sides, and the next minute we see her, a tangled mass of wreck and cordage, firmly embedded on the pitiless rocks. Don't suppose our fishermen had been quietly watching this and doing nothing to help. From the first, preparations had been made. Our friend, Jack, and a score of other active young men, had shoved off the only boat on the beach that had the faintest hope of living in a storm like this, and had been waiting in it close to the harbour mouth some minutes before the yacht struck. But so small was the chance of the frail boat living in such a sea, that many of the most experienced of the sailors made signals to prevent the men starting off to meet what they thought was certain death. Others thought it might be done, and waved contrary signals; and it was then that one saw what sort of women our sailor's wives are, for though many standing there with us had near and dear ones in that boat, and were suffering tortures of anxiety, not a word was spoken, but all was left for the men to do as they thought right.

As the yacht struck, a deep, wailing shout went up from all on land, and those in the boat knew what had happened, and the next moment we saw the boat plunge into the green waves at the harbour mouth. For a moment it seemed to stagger and quail, and then, impelled by those hands and muscles of iron, it was driven forward through the blinding spray into the angry sea beyond. Shall I ever forget how we watched that boat, now mounted high on the top of a wave, now for moments lost to sight, the men all straining at their oars to the utmost, and always creeping forward yard by yard? All this time, we on the Point could see, with increasing fears, that the hope of the yacht holding together till reached by the rescuers was but a faint one. Each monster wave that rolled in lifted it from the rocks and left it to fall back with an irresistible force midst spray and foam, that constantly wholly hid it from our sight; and even before the boat started, portions of the wreck were being tossed about on the sea, making its passage even more precarious. At one time a group of human beings was seen on the deck clinging to some cordage; but when the next wave passed, most of them had disappeared, and we knew they had perished before our eyes. It was difficult to distinguish objects midst the turmoil, but it soon was whispered among us that some one or more persons were crouching behind the bulwarks, probably lashed there for safety, and from an occasional flutter of a red scarf or garment, we feared there was an unfortunate woman among them; and once, as the waves receded from the deck, we distinctly saw a man rise up from the group and look for a moment towards the approaching boat, and then sink again beside his companions, just as the incoming wave swept high over the poor shelter the stout bulwark afforded.

If the yacht could only hold together a few minutes longer! But no! Once more it rises from its bed like some agonised, dying monster, and then as it falls back it parts in two, and half of it is a drifting mass of planks and timber, washing forward as if to meet the boat and destroy it. A portion yet remained fixed on the rock, and now and then we could still see the group crouching behind the bulwark. On and on fought the boat, now a little out of the direct line to avoid the wreckage, till it was close behind the wreck and partially sheltered by the rampart it formed against the sea; but at that moment all that remained of it was again lifted high in the air and dashed forward; and when the wave had passed by, there was only the frail boat with its brave crew to be seen on the surface. We see it pause for a moment, and then the oars all dip together, and the boat dashes forward. Someone leans over the bows, and there is a moment's struggle; but the mist and foam prevent our distinguishing clearly what is going on. After a while they evidently find there is nothing further that can be done; the boat is put before the waves and come dashing back towards land.

All on the Point hurried down to the entrance of the harbour; and many of the men, with coils of rope in their hands, stood ready to give assistance. As each wave rolled under the boat, it flew through the water, and then sank back again hidden from our sight; but nearer and nearer it come on, till at last on the crest of a wave it darted sharp round the Point, and lay tossing in comparatively calm water. Steadily its crew rowed it up the little harbour, and as it approached the beach scores of ready hands seized it and ran it high up on to dry land, and a cheer rang out above the roar of the wind to welcome those snatched from the jaws of death. But this was not responded to by the men in the boat. They all looked stern and anxious; and then we saw that Jack, who was crouched in the bows, was supporting in his arms the slight form of a fair young girl, with long, soft, tangled hair falling around her and forming a frame to the most beautiful saint-like face my eyes had ever seen. Her lips were parted in a smile, and her eyes looked down on a small boy about two years old, who was bound in her arms by a red scarf. At first I thought she was fainting or falling asleep, but the next moment—merciful Heavens!—I saw that the back of her sweet young head was battered in and bleeding, and that she was already beyond the storms of life and the cruel raging of the destroying elements.

Hard horny hands of rough women tenderly and deftly unwound the scarf from off the child; and Jack's wife, Mary, pressing him to her bosom, hastened with him to her cottage, while the fair dead form was carried to a fisherman's house close by, and a few days later was laid in its quiet grave in the old churchyard, within sound of the ruthless sea that had so cruelly beaten the young life out of it.

You may easily find the grave, for the fishermen out of their deep pity had a plain cross put over it, with just words 'Jack's mother' and the date of her death carved upon it. To this day, and I fancy for ever, the only name she will be know by is 'Jack's mother,' for all connected with that ill-fated yacht remains a mystery. Not a living creature escaped, except the frail little child. Many bodies were recovered during the next few days, and among them the remains of the man who had arrived the previous day in the dog-cart; but neither on any of the bodies, nor among the wreckage that came ashore, was anything found to lead to the identification of the yacht or its owners; and though the account of the disaster appeared in all the papers and was the talk of the county, yet no living soul has ever come forward to claim connection with the child or with any of those drowned.

It was thought at the time that the owner of the yacht was one of those desperate ruffians of Irish extraction that have from time to time arrived here from America, and that when he so hastily joined the vessel he was in fear of detection and was about to sail for America. Anyhow, the yacht was sighted by the gunboat sent to look after it, and chased and driven through the storm back to our little harbour, it being doubtless the intention of the fugitive to attempt his escape by land if he could once reach the shore. How miserable it ended you now know; but you don't know quite all, for I have not told you that, on reaching their cottage, Jack's wife found that the little one breathed. I have told you of the storm, and I have told you of the wreck; but words would fail to tell of all the love and care and attention that was bestowed for weeks—aye! For years, up to this day—on the little one. Only the recording angel can note such things, and only the God of love can reward them. Not that either Jack or his wife think of rewards either from earth or in heaven, for their love is wholly unselfish and all-satisfying; and were only the boy well and strong, I am sure that in all these realms there could not be found a more perfectly happy trio than Jack the fisherman, little Jack, and his adopted mother. Unfortunately it was discovered that in some way the child's back had been injured in the storm. For months he lay between life and death, at last to recover partially only in health, and without the use of his poor legs.

Many friends have come forward with help, and great London doctors have seen and attended the boy. Till lately they gave little hope, but, thank God, there has been during the past year a slow but steady improvement, and they now think in time the boy may grow strong in health, but there is no hope of his ever walking without his crutches.

Fortunately nature has bestowed many gifts on the poor child that compensate him somewhat for his loss—first, an intensely loving, unselfish nature; and secondly, a perfect voice and passionate love of music. Already he is carried each Sunday to church by his father, and his voice in the choir is celebrated for many miles round, and has so impressed the organist at the cathedral at Marshford[44] that he either comes himself, or sends one of his pupils, to give the boy a lesson once a week, and there is not a better violinist within the bounds of this county than our little Jack is. His father is so proud of the boy's gifts that I have know him, when wind-bound in a harbour down the coast twenty miles away, walk over the whole distance on a Sunday morning and back at night rather than miss carrying the little fellow to church and hearing him sing there. But it is eleven o'clock, and we were up all last night. What, no grog? Well, good night! Come and see me when you can, and come and watch the sea with me in another storm, and we will see if I can't rake up another story of the doings of the rough heroes of our neighbourhood who go down to the sea in ships. Good night, good night!

And so one of the pleasantest evenings I had spent for a long while was over.

Oh, dear! Oh, dear! What a muddle, what a hodge-podge I have made of this pen work! I sat down thinking it would be quite easy to write a book on "Rat-catching for for the Use of Schools," and I have drifted off the line here, toppled into a story there, and been as wild and erratic in my goings on as even Pepper would be with a dozen rats loose together in a thick hedge. Well, I can't help it. I am not much good at books, and it ain't of much consequence, for during the last few days I have heard from half a dozen head-masters of schools that they find

---

[44] This appears to be a fictional location, as I can find no modern evidence for a town of that name in England.

the art of rat-catching is so distasteful to their scholars, and so much above their intellect, and so fatiguing an exercise to the youthful mind, that they feel obliged to abandon the study of it and replace it once more by those easier and pleasanter subjects, *Latin and Greek.* Well, I am sorry for it, very sorry. I had hoped to have opened up a great career to many young gentlemen, but have failed; and I can only console myself with thinking that one can't make silk purses out of—you know what. Mind, in this quotation I am not thinking of myself and my failure.

.